To …

Magda and Mohamed

Catherine

Mary, Tom and Charles

Investment Decisions in Advanced Manufacturing Technology

A fuzzy set theory approach

MAGDY G. ABDEL-KADER
School of Business and Economics, University of Exeter

DAVID DUGDALE
Bristol Business School, University of the West of England, Bristol

PETER TAYLOR
Bristol Business School, University of the West of England, Bristol

Routledge
Taylor & Francis Group
LONDON AND NEW YORK

First published 1998 by Ashgate Publishing

Reissued 2018 by Routledge
2 Park Square, Milton Park, Abingdon, Oxon, OX14 4RN
711 Third Avenue, New York, NY 10017, USA

Routledge is an imprint of the Taylor & Francis Group, an informa business

A Library of Congress record exists under LC control number: 98073748

ISBN 13: 978-1-138-32949-2 (hbk)
ISBN 13: 978-0-429-44814-0 (ebk)

Contents

List of Figures

List of Tables

Preface

This book is the outcome of over four years work carried out, principally, by Magdy Abdel-Kader whilst studying for his Doctor of Philosophy at the University of the West of England, Bristol. Thanks are due to the Egyptian Government for the sponsorship Magdy received during his studies and to the University of Cairo which gave Magdy leave of absence during this period.

It was intended that the research should lead to an operational model of AMT decision making that incorporated the mathematics of fuzzy set theory, and this aim was achieved. Magdy's interest in this field was stimulated early in his PhD studies by a period of work in the United States under the supervision of Professor Hamid Parsaei of the University of Louisville, an expert in the application of fuzzy set techniques in advanced technology decisions. Thanks are due to Professor Parsaei for his help during the early stages of Magdy's research.

On joining the University of the West of England in February 1995, Magdy transferred supervision of his work to Peter Taylor and myself, and began his empirical work in earnest. A review of the AMT decision making literature led Magdy to the conclusion that the existing empirical literature did not sufficiently describe the manner in which these decisions were taken in practice, and this naturally led him to consider further work in this area. Two separate approaches were adopted, the first based on a questionnaire survey, and the second based on semi-structured interviews. Triangulation of research methods was adopted in order to minimise the well known weaknesses of either method when used solely.

The research design was established by the three members of the research team, and the field study interviews were carried out by Magdy Abdel-Kader and myself. The authors believe that these studies add considerably to understanding of the important strategic decisions regarding AMT. The particular design of the studies facilitated the relatively sophisticated statistical analysis, and provided insights into the practical issues that surround major investment decisions. Statistical analysis was possible because companies responding to the questionnaire survey could be divided into three categories - those not investing in AMT, those with partly integrated AMT systems and those with fully integrated systems. This part of the research, set out in 'scientific' format and focusing on the testing of a series of hypotheses, has been published in the journal *Management Accounting Research* (see Vol. 9, pp 261-284), and the authors are grateful to the publishers of the journal for permission to publish a version of the study here.

The questionnaire survey and the field study together provided a convincing description of the key features of AMT investment decision making, and this was taken as the basis for further work. This consisted of the development of an empirically grounded model which allows decision makers to use numerical approximations and/or ambiguous or vague concepts in specifying the input data. This is achieved by the use of fuzzy set theory. Through this work the existing literature has been extended by developing a model which can accept *both* numeric and linguistic data, and by proposing a novel method for ranking fuzzy numbers.

Although the authors have all made contributions to the work, the contribution of Magdy Abdel-Kader is paramount. The support of Dr John Pointon of the University of Plymouth must also be acknowledged. John has encouraged us in the preparation of this work, and his help in some of the more mathematical sections is appreciated. The authors also acknowledge the helpful advice of participants at the British Accounting Association conferences held in Plymouth (1995), London (1995), Cheltenham (1996), Birmingham (1997) and Manchester (1998), and the comments of anonymous referees. Despite the efforts of these colleagues, any errors that remain in the text, are of course, the sole responsibility of the authors.

Dr David Dugdale
Bristol Business School
University of the West of England, Bristol
September, 1998

List of Abbreviations

AGVS	Automated guided vehicles system
AHM	Automated material handling
AHP	Analytic hierarchy process
AMT	Advanced manufacturing technologies
AS/RS	Automated storage and retrieval system
CAD	Computer aided design
CAM	Computer aided manufacturing
CAPM	Capital asset pricing model
CNC	Computer numerical control
Com	Competitiveness
DCF	Discounted cash flow
ES	Existing system
FIRR	Fuzzy internal rate of return
Flex	Flexibility
FMS	Flexible manufacturing systems
FMT	Flexible manufacturing technology
FNFM	Fuzzy non-financial measure
FNPV	Fuzzy net present value
FPB	Fuzzy payback
FRM	Fuzzy risk measure
FROI	Fuzzy return on investment
IRR	Internal rate of return
LS	Linguistic scale
NPV	Net present value
PB	Payback period
R&D	Research and development
RI	Random index
ROI	Return on investment

1 Introduction

Recent years have seen rapid changes in manufacturing, a most important sector in many economies. There is a clear trend from mass production of standard products to the production of a greater variety of custom-made products with shorter product life-cycles, as well as increasing complexity and sophistication of product design. Consumers now demand greater variety, better quality, reliability, and lower prices. To survive, manufacturing firms need to reduce production costs for small batch sizes and greater product-mix complexity, while producing consistently better quality products. Firms must also be able to introduce new products quickly, and cope with shorter delivery cycles (Naik and Chakravarty, 1992). Advanced Manufacturing Technologies (AMT) such as flexible manufacturing systems and robots help to achieve this.

Major investments such as those in advanced manufacturing systems have a significant impact on the long-term performance of the company as a whole. A study conducted by the Institute of Production Engineers (1987) concluded that investments in advanced manufacturing systems should be based on a carefully prepared manufacturing strategy which should form an integral part of overall company strategic planning.

Consider an overall strategy that attempts to achieve a broad range of differentiated products. The company seeks to differentiate its products across the complete range, looking to add unique benefits such as better quality or greater convenience. Here, manufacturing strategy might aim to acquire an advanced manufacturing system that enables the company to produce a variety of top-quality products. This may then lead the company to re-consider other strategic issues such as rationalisation or integration of facilities and improvement of workflow. Consequently, investments in AMT need to be considered as part of the overall corporate strategy, and not simply as a series of incremental capital budgeting decisions (Schroeder, 1989, p. 268).

Advocates of 'strategic' AMT investment have fiercely criticised traditional methods of financial evaluation which, they argue, unfairly disadvantage AMT projects. The flavour of these criticisms is nicely captured in the title of a paper by Robert Kaplan (1986): 'Must CIM be justified by faith alone?' The main theme of this book is the conflict between 'financial' and 'strategic' investment evaluation and efforts to synthesise the two.

This chapter begins with a review of advanced manufacturing technologies and the strategic benefits they can bring, then we turn to the criticisms of traditional investment evaluation methods.

Advanced Manufacturing Technologies

Rapid development in technology has made available a host of automated devices with diverse applications in manufacturing industries. Functional areas such as product design, product testing, process design, fabrication, spray painting, assembly, inspection, material handling, warehousing, and production planning and control have been dramatically affected by automation (Huang et al., 1991). Generally, automation refers to the use of mechanical devices controlled by a host computer that can be programmed and reprogrammed with various operating instructions.

There are many forms of AMT such as Computer Numerical Control (CNC), Robots, Automated Material Handling (AMH), Computer Aided Design (CAD), Computer Aided Manufacturing (CAM), Flexible Manufacturing Systems (FMS), and Computer Integrated Manufacturing (CIM). Following Meredith and Suresh (1986), these forms can be categorised as 'stand alone systems' (CNC and robotics), 'partly integrated systems' (AMH, CAD and CAM), and 'fully integrated systems' (FMS and CIM).

Computer Numerical Control

The basic components of an automated manufacturing system are computer numerically controlled (CNC) machine tools. A conventional machine tool, such as a turret lathe, a milling machine, or a grinder, is controlled by an operator who determines the cutting speed of the machine, the depth of cut, the initial and successive locations of the cutting tool and its movement (or alternatively that of the workpiece) to achieve the required dimension and geometry. In a CNC machine tool, the above factors are determined and controlled by a computer according to a set of instructions (part program) prepared for the product being processed. These instructions contain information necessary for machining the product, such as the movement of the tool or the workpiece specified by successive co-ordinates, cutting speed, depth of cut, etc. Each CNC machine is individually controlled by a single computer dedicated to that machine (Hilton, 1991).

Robots

The Robotics Institute of America (RIA) defined a robot as:

> 'A robot is a programmable, multi-functional manipulator designed to move material, parts, tools, or specialised devices through various programmed motions for the performance of a variety of tasks'. (Asfahl, 1992, p. 132)

Groover et al. provided a more general definition of robotics:

> 'An industrial robot is a general-purpose, programmable machine which possesses certain anthropomorphic, or humanlike, characteristics'. (1986, p. 5)

The word 'programmable' in the definitions is very important because the functions of a robot are controlled by a programmable microprocessor-based system that can be linked into a computerised work environment. Not only can it process work automatically but it can also react to a changing work environment, receiving and transmitting data relating to its function. Therefore robotics are an essential part of any computer integrated manufacturing system.

In the past, robotics have been mainly used in stand-alone activities and in the early days, they performed hot, heavy and hazardous jobs previously performed by people. In addition to these socially desirable functions a robot, once programmed, can work at the same rate continuously throughout the day without losing quality. But today the emphasis is changing. Improved productivity and quality, rather than the displacement of people, are the driving forces behind the introduction of robotics in the workplace (Kochan and Cowan, 1986).

Automated Material Handling

Automated Material Handling (AMH) comprises Automated Storage and Retrieval System (AS/RS) and Automated Guided Vehicles System

(AGVS). AS/RS is a computer-controlled stocking system in which items are stored on racks and received and retrieved using computerised robotics, cranes, and/or similar devices. AGVS provides unmanned transportation of materials in the factory and is equipped with automatic guidance devices programmed to follow certain paths.

Koening (1990) suggests that, in flexible manufacturing systems, an AGVS can be loaded by the AS/RS and material delivered via flexible pathways to any one of a number of machining centres.

Computer Aided Design

Computer Aided Design (CAD) is computer software used by engineers in the design of products or production processes. It helps to create or modify engineering designs and stores the information in a computer database. In these applications, the operator constructs a highly detailed drawing on-line using a variety of interactive devices and programming techniques. Facilities are available for replicating basic figures; achieving exact size and placement of components; making lines of specified length, width, or angle to previously defined lines; satisfying varying geometric and topological constraints amongst components of the drawing etc. (Kochan and Cowan, 1986).

Computer Aided Manufacturing

Computer Aided Manufacturing (CAM) is the term commonly used for all computer-controlled activities that are involved in turning raw material into finished products.

Flexible Manufacturing Systems

The concept of Flexible Manufacturing Systems (FMS) was developed in the late 1960s and early 1970s. There is growing interest in the development and implementation of FMS and, to date, hundreds of these systems have been implemented around the world, with Japan leading in terms of the number of applications and associated management and organisational success (Kusiak, 1985).

The United States National Bureau of Standards defines FMS as:

'An arrangement of machines (usually numerically controlled machining centres) with tool changers interconnected by a transport system. The transporter carries work to the machines on pallets or other interface units so that work-machine registration is accurate, rapid and automatic. A central computer controls both machines and transport system. Flexible manufacturing systems sometimes process several different workpieces at any one time'. (Nagarur, 1992. p. 799)

The Charles Stark Draper Laboratory defines a FMS as:

'Computer-controlled configuration of semi-independent work stations and a material handling system design to efficiently manufacture more than one kind of part at low to medium volumes'. (1984, p. 19)

It can be seen that an FMS consists of three components: work stations (mostly numerically controlled machine tools), an automated material handling system, and a network of supervisory computers and programmable logic controllers.

In FMS environments, parts are brought automatically on pallets to input stations, the material handling system allows the jobs to move from the input stations to any work or storage station, and computers process instructions, ensuring that necessary job tools are available (Medearis et al., 1990). Computers perform several functions in these systems: scheduling and monitoring operations, handling material control, and taking appropriate actions in case of sudden changes in the system.

Industrial Robots in FMS If computers are the 'brains' of FMS, robots are often the 'hands' performing a range of tasks. Robots for handling typically transfer objects from one place to another and load/unload machine tools with parts and tools. Tool-Operating Robots find wide application, for example in coating (painting, underbody protection), spot welding, machining, drilling and trimming. Assembly Robots are used to transfer and fit parts, and are usually designed to cope with variation of parts, variation in the orientation of parts, defective parts and different products assembled on the same flexible assembly line.

Computers, robots and other automated devices can offer great flexibility allowing the system to quickly adjust to any changes in relevant factors like product, process, loads and machine failures. With computer integration, it may even be possible to equip the system with a certain degree of self-diagnostic and adaptive control ability.

Computer Integrated Manufacturing

Koening (1990) sets out seven steps which in a full computer integrated manufacturing (CIM) system would be linked:

1. Obtain product specifications.
2. Design method for producing the product, including design and purchase of equipment and production processes.
3. Construct a production schedule.
4. Purchase raw materials according to the schedule.
5. Produce in the factory.
6. Monitor quality, technical compliance and cost control.
7. Ship the completed product to the customer.

In CIM all NC machines are controlled by a central computer, which also controls movement of parts by the automatic material handling system. This material handling system connects NC machines to each other as well as to the loading and unloading areas. Parts are attached to standard size pallets, which facilitates not only the movement on to the transportation system, but also the mounting of parts on to the NC machines. The dual purpose use of the pallets saves time as part locating jigs and fixtures are not necessary. Various types of automatic material handling systems may be used, including power roller conveyors, shuttle conveyors, automatic guided vehicles and robotics. In addition, the central computer is also used for process design, planning, monitoring and control, shop floor controls, inventory control, and preparation of performance evaluation reports. It can also be used to direct and monitor automatic quality control checks and tests (Dhavale, 1989). In short, when all activities in the factory are linked through a central computer, the factory is considered to have CIM.

Benefits of AMT

The decision to invest in AMT may be motivated by a variety of reasons. O'Brien and Smith (1993) identified the following:

1. Adding capacity to meet increased demand and/or product variety.
2. Meeting the need for constant replacement of old and obsolete technology. Rarely is such new investment an exact replica of the old, and considerable adaptation may be required if the full potential of new technologies is to be exploited.
3. Exploiting wider opportunities in product and process innovation.
4. Obtaining major strategic benefits by improving the competitive advantage of an organisation. Investment could, for example, open up new markets for a company or be the only means of surviving in existing markets. Shortening the time to market, optimising the use of available skills, improving quality, increasing product life through innovation and promoting flexibility in products and processes are all areas which may enhance an organisation's ability to deal with uncertainty and hence survive in an increasingly dynamic and competitive environment.
5. Although innovation and product quality are essential in securing market share, no company can hope to remain competitive without continued attention to the reduction of manufacturing costs. The trend toward "lean production" requires attack on costs in every area of a company's activities.

AMT can provide a range of benefits. The following are commonly cited:

1. Reduced direct labour costs. Direct labour costs are generally reduced when automated equipment is purchased. However, this reduction is rarely sufficient to justify automation.
2. Reduced inventory costs. Automated equipment is more reliable, more consistent, and faster than traditional equipment. Therefore, it can reduce inventory costs by reducing the quantity of inventory on hand, and by releasing space for other uses.
3. Reduced quality costs. Due to greater reliability and consistency of output, automation results in less defects and less waste, scrap, and rework costs. In turn, reduction in defects and related problems lead to reduction in warranty expenses. Also, greater product uniformity and reliability mean that fewer inspections are needed.
4. Faster throughput time. The greater efficiency of an automated process will decrease the production throughput time and can therefore increase the total output for a given period.

5. Increased manufacturing flexibility. Set-up time can be reduced through automation, thereby increasing manufacturing flexibility. Also, the flexibility of automated equipment usually means a longer life than traditional equipment.
6. Faster response to market changes. Due to the flexibility of automated equipment, the firm can respond more quickly to changes in customer tastes and needs.
7. Increased learning effects. Automating a facility or a process is difficult, both in a technical and an operational sense. Firms that hold back, fearing the complexities of automation, soon fall behind their competitors in recognising and in being able to utilise the newer technology as it comes onto the market.
8. Avoiding capital decay. Capital decay can be defined as a loss in market share resulting from technologically obsolete products and operations. Retention of market share, or even an increase in market share, is perhaps the most significant intangible benefit that can be gained from automation. Some firms reject automation after assuming that sales will remain unchanged even if they do not automate. However, if a more efficient process is available, a competitor undoubtedly will invest in it, thereby gaining a competitive advantage (O'Brien and Smith, 1993). The appropriate assumption is that if a firm fails to take advantage of a new technology, it will not be able to maintain the status quo in terms of sales; rather, sales are likely to decline.
9. Higher quality output. Automation allows a higher quality of output that can greatly strengthen the image and competitive position of a company. Higher quality promotes confidence on the part of customers and improved reliability that can provide access to expanding, world-wide markets.

The major difficulty in justifying AMT investment is in accounting for all the benefits which should be gained. Some benefits such as direct labour savings, direct material savings, less frequent set-ups, higher quality output, and the savings from reduced material and in-process inventories can be quantified. However, a second category of more intangible benefits such as: greater manufacturing flexibility, learning effects, and improved employee morale cannot be easily quantified. In the following chapters we shall evaluate the suggestions that have been made to overcome this problem and develop a an appropriate model for investment decision making regarding AMT.

Traditional Investment Appraisal

Traditional investment models based on return on investment (ROI), payback period (PB) and discounted cash flow (DCF) methods are widely used to evaluate potential investment projects in practice (Somers and Gupta, 1991; Wilner et al., 1992; Primrose et al., 1985; Primrose and Leonard, 1986a; and Kulatilaka, 1984). Theoretically, the discounting methods: net present value (NPV) and internal rate of return (IRR), are preferred, and in addition, a US survey (Wilner et al., 1992) reported that a large percentage (70%) of firms specified one of the DCF methods as the primary technique for the evaluation of high technology investments.

To use DCF methods three estimates are required (Chan and Lynn, 1993): a forecast of annual net cash flows, a discount rate (theoretically, the firm's cost of capital), and an estimate of the life of the project. The firm should invest if the NPV of the estimated cash flows is positive or, alternatively, if the IRR exceeds the firm's cost of capital.

The application of traditional appraisal methods to AMT projects usually follows a standard procedure based on the identification of the relevant cash flows, which are the incremental cash flows of the project if adopted. As the subsequent analysis can only be as good as the data on which it is based, every effort should be made to estimate the cash flows as accurately as possible. While projected benefits will often be subjective (and quantification of some benefits may prove impossible), it should be possible to establish fairly accurate cost estimates for the project. We briefly review some of the issues which ought to be considered.

Acquisition (or Initial) Costs

The decision to purchase any form of AMT is different from an investment in conventional technology. AMT cannot be bought off the shelf and plugged into the factory. The system (hardware and software) may be bought from machine tool manufacturers or from vendors, and the interfacing may be designed in-house. The final cost will thus exceed the initial purchase cost due to the costs of interfacing, rearrangement, and modification (Kulatilaka, 1984). In new plants that are *designed* around advanced technology, these costs will be less.

One important factor that may be overlooked is the opportunity cost of existing assets or resources. For example, the opportunity cost that may arise from using an asset which is already owned by the firm in the new system, or from the time lost during the process of transformation from the conventional manufacturing system to the advanced one. Such costs should be considered in the evaluation process.

Annual Operating (or Running) Costs

The operating costs of AMT include all the costs which the system needs to run effectively. These costs mainly consist of factory costs (which include direct material, direct labour, and overhead), and selling and administrative expenses.

Klahorst (1986) indicates that for the economic evaluation of FMS there are 14 cost factors that should be considered:

1. Direct labour costs.
2. Material set-up costs.
3. Tooling costs.
4. Materials handling costs.
5. Part inspection costs.
6. Equipment maintenance costs.
7. Shop supervision costs.
8. Production control costs.
9. Manufacturing engineering costs.
10. Plant facility costs.
11. Inventory costs.
12. Fixturing costs.
13. Prototype and new part costs.
14. Rework and scrap costs.

Investment Decisions for AMT

Although consideration of the acquisition of AMT is an investment decision, AMT systems have special characteristics that make the evaluation process more complex than for conventional types of manufacturing systems. First, advanced systems are capital intensive. They require a huge amount of investment in all stages of implementation - planning, purchasing, installing and operating- even when only a few items are purchased. Second, advanced systems are much more flexible than conventional systems. This flexibility maintains the value of the equipment over the long run, reducing its rate of depreciation. However, the advantages of such flexibility are not easily quantified in traditional appraisal methods. Third, there is synergy when advanced technologies are linked together. Users consistently report qualitative benefits from such linked systems, such as fast response to future customers needs that are deemed more important than the normal cost savings (Meredith and Suresh, 1986 and Garrison, 1991). Finally, lack of experience of high technology systems can make them risky decisions.

The specific nature of AMT projects has caused many authors to criticise the application of traditional investment models in evaluating AMT investments (see, for example, Medearis et al., 1990 and Mensah and Miranti, 1989). The criticisms are of two main kinds. First, traditional investment models emphasise quantitative, financial analysis, but it is argued, fail to capture many of the 'intangible' benefits which should flow from AMT investment. Second, traditional models may militate against AMT investments through high discount rates and short payback targets which systematically penalise long-term investments. Recognition of these issues has led a number of researchers to conclude that traditional investment models should not be used to evaluate investment in AMT.

A major problem when applying traditional investment models to AMT decisions is the failure of these models to quantify all of the benefits of the acquisition of automation. Traditional models focus only on future net cost savings such as reduced labour and energy costs which are easily quantifiable, and they ignore the other benefits such as improved product quality, greater manufacturing flexibility, learning effects, effects on employee morale, and short lead times. Consequently, these models lead to investment myopia, in particular, creating a bias against 'strategic' investment in AMT. This problem is stressed in almost all of the relevant literature (see, for example, Chen and Small, 1996; Accola, 1994; Cheung, 1993; Lavelle and Liggett, 1992; Naik and Chakravarty, 1992; Azzone et al., 1992; Rayburn, 1989; Park and Son, 1988; Srinivasan and Millen, 1986; Kaplan, 1986; ACARD, 1983; and Knott and Getto, 1982).

However, Primrose (1991) dissents from this broad consensus. He denies the existence of the quantification problem claiming that all the benefits of AMT are quantifiable in cash flows through their effect on current sales and costs (see also, Primrose and Leonard, 1986a; 1986b; 1987). This may be true for some benefits such as reduced floor space, reduced inventories and reduction in rework, inspection and warranty costs, but quantifying benefits such as flexibility, learning effects, reduced lead times, employee morale or quality is, at best, a complex task, and, at worst, almost impossible to undertake. Arguably, these benefits are most critical in evaluating the decision of whether to acquire AMT.

The problems inherent in quantifying AMT benefits can be illustrated by trying to estimate the cash flows from increased product quality. Four relevant dimensions of product quality can be identified: product durability, product reliability, product aesthetics, and product performance. These attributes, in turn, depend on three factors: the product design, the materials used in making the components, and the engineering specifications and tolerances observed in product manufacture. The trade-off between these factors and their effects on product costs and market niche has to be assessed as a part of organisational strategy. Also, the savings resulting from reduced raw material and in-process inventories are not easy to estimate because the relationship between organisational costs and the level of inventories is not well-defined. Additionally, these costs vary with the volume of discrete events such as the frequency of set-ups and the number of different types of products, and not with the volume of output. Under these circumstances, cash flow estimates depend on expected future management action with respect to organisational costs (Mensah and Miranti, 1989).

Meredith and Suresh (1986) argue that evaluating AMT investments on the basis of traditional investment models represents a major roadblock to the automation of many factories. There is very little doubt that non-investment in AMT can, in many cases, have serious long-term consequences and that innovation and change are part of our daily business life (Lefley, 1996). Companies failing to recognise the importance of taking opportunities provided by technological improvement in manufacturing can risk their continued profitability and long-term survival (ACARD, 1983; Attaran, 1992; Lefley, 1996; and Chen and Small, 1996). Lefley (1996) summarises the problem as follows:

> '... Management is placed in a dilemma in that on the one hand they wish to invest in new technology, such as AMT, but on the other hand they find it difficult to justify the capital expenditure using traditional financial appraisal techniques. The conventional financial evaluation methods are well-established, well documented, while the methodologies for evaluation of strategic, intangible benefits ... are less formalized and less understood. Evidence suggests that there is a need for a more sophisticated approach to the appraisal of AMT projects, one that will take into account the strategic nature and the full benefits from such investments'. (p. 347)

The need for a more sophisticated approach for the evaluation of AMT investment was also a conclusion of an earlier empirical study conducted by Currie (1994). She stated that:

> '... The study concludes by suggesting that new methods of evaluating AMT should be developed which include a wider array of financial and non-financial benefits. This would improve managements' understanding of some of the key advantages of AMT and, in the process, supplement traditional management accounting techniques (DCF, NPV, payback) by considering the benefits of quality, organisational learning, training and process improvement and innovation'. (p. viii)

The implications of this issue are clear. While DCF methods may be economically sound, their application is often deficient in relation to AMT investment because major qualitative variables are ignored. As both Lefley (1996) and Currie (1994) suggest, new, more sophisticated methods of AMT investment appraisal are needed, extending traditional DCF methods so that *all* costs and benefits associated with AMT proposals are properly evaluated. We begin our discussion of such models in chapter 2.

A further problem in evaluating AMT investments is how to consider a project's risk in the investment decision model. Strictly, there is a difference between the terms 'risk' and 'uncertainty'. The term risk should be used when it is possible to assign objective probabilities while uncertainty implies that it is not possible to assign such probabilities (see, for example, Pike and Dobbins, 1986). However, the term risk is used here to refer to the uncertainty of obtaining the expected benefits (tangible or intangible) from an investment project.

One method often used to account for the high risk inherent in new technologies is the use of an arbitrarily high hurdle discount rate (see, for example, Accola, 1994; Canada and Sullivan, 1990; Kaplan and Atkinson, 1989; and Kaplan, 1986). This method suffers from two main criticisms. First, arbitrarily high hurdle thresholds favour short-term projects over long-term projects which have large cash flows in the latter part of their lives. This is because the discount rate compounds geometrically each time period, so cash flows received five or more years in the future will be penalised severely in the analysis. Also, as pointed out by Kaplan and Atkinson (1989), much of the risk from new technologies will probably be resolved early in the project's life. If there is uncertainty as to whether a new piece of equipment or new technology will work, the outcome will be known in the first or second year. Second, Accola (1994) and Ronen and Sorter (1972) objected to this approach on theoretical grounds because it uses a single measure to reflect both expected cash flows and the riskiness of the expected cash flows. As there are many different determinants of risk, it is difficult to capture all aspects of a project's riskiness through a single modification of the discount rate. Also, adjustments to the discount rate are affected by managers' attitudes toward risk rather than by an explicit representation of the risks inherent in the investment alternatives.

Outline of the Text

The remainder of the text sets out the issues in AMT investment decision making and recommends a new approach that can take account of *all* the decision factors and permits the estimation of input variables in terms such as 'about £2 million' or 'product quality is important'. Often in practice, the inputs to the decision model are specified in exact terms and the inherent vagueness in the basic data is ignored (and forgotten!) in subsequent analysis.

In chapter 2, previous AMT decision models are reviewed. Typically, these are based on multi-attribute scoring and lead to an overall 'score' for each alternative under consideration. The models range from the simple weighting of individual attributes to the relatively sophisticated analytic hierarchy process which allows sub-dimensions to be specified for each decision attribute. We conclude that these models suffer from two major drawbacks. First, there is scant justification for the particular models chosen or for the variables included (and excluded) in the decision process. Second, most of the models require exact estimates of the input data.

The lack of justification for the extant models is addressed in chapters 4 and 5. Chapter 4 reports the findings of a questionnaire survey of major UK manufacturing companies specifically designed to identify the key issues that must be addressed when evaluating AMT projects. Chapter 5 reports the results of a field study of nine UK companies. A series of AMT cases are reported, insights drawn and a number of general conclusions reached. The questionnaire survey and field study support each other and allow the development of an AMT investment model which is empirically grounded and which should therefore find favour with practitioners.

Chapter 6 introduces fuzzy set theory as a means of handling approximate or vague estimates. Much of this chapter is a distillation of the existing literature. However, the analysis reveals some difficulties with existing methods, especially in relation to the ranking of fuzzy numbers. This is a particularly important issue when choosing between competing investment projects and, therefore, a new ranking method is proposed which may be an improvement on previous suggestions.

Chapter 7 develops a model for AMT investment decision making based on the pre-existing descriptive literature and the results of empirical studies reported in chapters 4 and 5. The analytic hierarchy process and fuzzy set theory provide the mathematical tools which form the basis of the model. The model represents a synthesis of several previous models and the authors' empirical work and is designed to address both theoretical issues (such as combining both quantitative and qualitative variables) and the incorporation of variables and features which make the model credible to practitioners.

Chapter 8 provides an overview of the book as a whole, drawing together a number of themes and suggesting possible directions for future research.

2 A Review of Previous Models for the Evaluation of Investment Decisions in AMT

Two major problems in applying traditional models for evaluating AMT investments have been discussed in the previous chapter: the problem of quantifying the intangible benefits of AMT investments, and the problem of measuring and incorporating project risk into the investment model. These problems have led to a growing literature devoted to investigating them and providing possible solutions. This literature can be divided into two major strands. In the first strand researchers attempt to develop theoretical models which aim to overcome the shortcomings of traditional investment decision models based on ROI and cash flow analysis. This type of research can be described as 'normative'. The second strand of literature is based on empirical investigation of the practice of AMT investment decision making by either survey based research or field studies. This type of research can be described as empirical or 'descriptive' research. The major difference between these two strands is that normative researchers seek to formulate decision models that describe *how decisions should be made* while descriptive researchers seek to explore the practice of decision models *as it exists* (see, for example, Ijiri, 1972; Bell et al., 1988; and Ryan et al., 1992). Bell et al. (1988) add a third type of decision making research which they call 'prescriptive' research. They use this term to describe a type of research which combines both normative and descriptive approaches in producing a decision model.

The extensive literature includes numerous normative (theoretical) models specifying how AMT investment decisions ought to be made, but few descriptive (empirical) studies which describe AMT investment decision making practice and no prescriptive research studies. In subsequent chapters an empirically grounded normative model for evaluating AMT investments is developed thus contributing to the 'prescriptive' AMT investment decision making literature.

In this chapter 'normative' models of AMT investment decision making are reviewed. Empirical studies are reviewed in the next chapter.

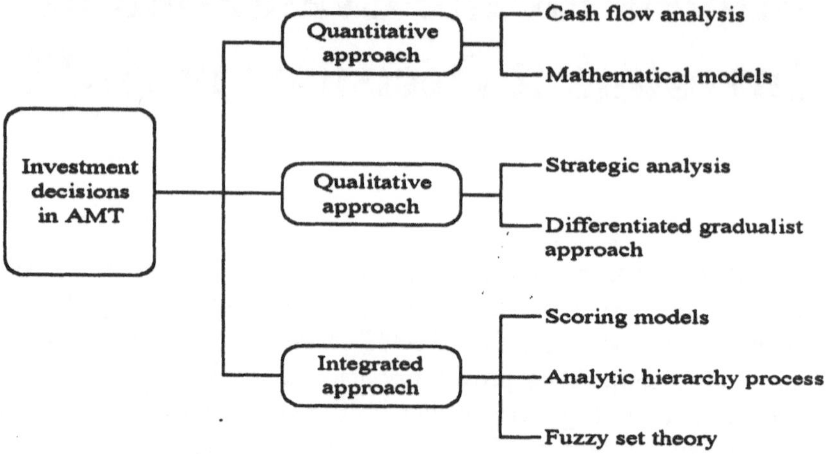

Figure 2.1 A proposed classification scheme

Much of the literature of AMT investment decision making emphasises the problem of quantifying intangible benefits. The literature can be classified into three approaches: quantitative, qualitative and integrated[1], see figure 2.1. Each of these will be considered in turn.

The Quantitative Approach

This approach focuses mainly on easily quantifiable variables. Traditional financial appraisal methods are considered in the next section, and then various mathematical models are reviewed.

Cash Flow Analysis

The most common cash flow based analyses are payback calculations and discounting methods (net present value and internal rate of return)[2]. These methods are widely used. For example, a survey conducted by Klammer et al. (1991) showed that 42% of one hundred US large industrial firms used the payback method as a primary or secondary technique and the discounted cash flow methods, IRR and NPV were used by 57% and 46% of firms respectively in the evaluation of high technology investments.

Payback Period Method The payback period is the time required to recover the initial investment outlay. The project is acceptable if the payback period is equal to or less than a predetermined payback period. The method assumes that the project with the shortest recovery period is the best/lowest risk project. The payback period method has many advantages and is much used in practice. The advantages include: quick and simple to calculate; needs minimal information for computations (Randhawa and West, 1992); and no need to forecast cash flows over the whole project life (Lumby, 1995, pp. 42-43).

However, the payback period method has been criticised for its use in evaluation of investments in general and particularly investments in AMT because: it is short-term as it ignores cash flows after the payback point, and hence is biased against investments with higher returns in their later years, and is inadequate for rigorous analysis of all the variables and for systematic comparison of competing projects.

Primrose and Leonard (1986) also drew attention to the difficulty of determining the required payback period. They considered the relationship between IRR and payback period using a hypothetical example, and showed that companies using a 2 year payback period require an after-tax IRR of 49%, and companies using a 3 year payback period require an after-tax IRR of 31% (see figure 2.2).

The extremely high IRRs calculated were due to two defects in the payback method. First, cash flows after the payback period are ignored, and, second, the time value of money is not taken into account. The latter defect can be overcome by using discounted cash flows instead of 'pure' (non-discounted) cash flows. For example, assume the estimated cash flows of a project during its 5 year life are:

Year	Cash flows (£)
0	(300,000)
1	100,000
2	120,000
3	120,000
4	90,000
5	50,000

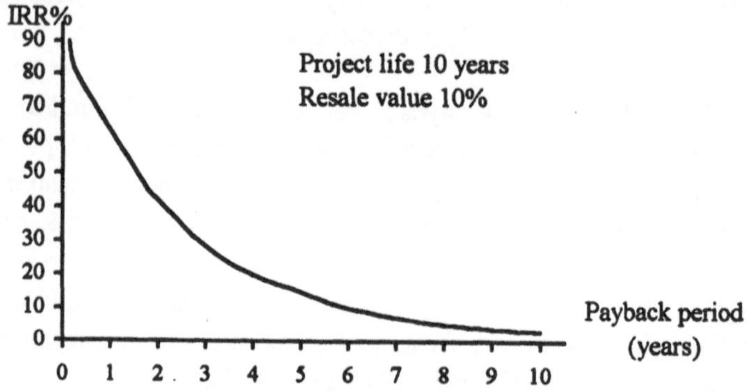

Figure 2.2 A relationship between IRR and Payback period

Source: Adapted from Primrose and Leonard, 1986a, p. 290.

Assuming the cash flows are realised at the end of each year, the payback period is 3 years. However, assuming that 10% is an appropriate discount rate, the discounted payback period is computed as follows:
Discounted cash flows:

Year	Cash flows	× Discount factor	=	Present Value
0	(300,000)	× 1	=	(300,000)
1	100,000	× 0.909	=	90,900
2	120,000	× 0.826	=	99,120
3	120,000	× 0.751	=	90,120
4	90,000	× 0.683	=	61,470
5	50,000	× 0.621	=	31,050

The discounted payback period is 4 years as compared against 3 years for simple payback.

DCF Methods Net present value (NPV) and internal rate of return (IRR) are both discounted cash flow (DCF) methods. The NPV is the figure that represents the sum of future cash flows (positive or negative) discounted to time zero (equation 2.1).

$$NPV_p = \sum_{t=0}^{n} \frac{CF_t}{(1+r)^t} \tag{2.1}$$

Where:

NPV_p: the net present value of project p,
CF_t : the cash flows of the project in time t $t = 0,1,\dots\dots\dots n$,
r : the discount rate, and
n : the life of the project in years.

The decision rule for choosing among competitive alternatives is that the one with the greatest NPV is the most desirable. On the other hand, the IRR is the discount rate which makes the sum of discounted cash flows equal to zero (equation 2.2). In equation 2.2, r is the IRR, usually computed through a number of trials. The decision rule for IRR is similar to that for NPV, namely, the alternative with the highest IRR is the most preferable.

$$\sum_{t=0}^{n} \frac{CF_t}{(1+r)^t} = 0 \tag{2.2}$$

DCF methods are widely recommended for the evaluation of AMT investment decisions.

Kulatilaka (1984) advocated the use of NPV to evaluate investment in automation projects. Three groups of cash flows were considered in the analysis: first, those related to initial costs; second, those related to operating costs and benefits; third, those related to tax effects. He used the Capital Asset Pricing Model (CAPM) to arrive at a discount rate that incorporated the systematic risk (beta) associated with the different components of cash flows. His main contribution was the suggestion that the project beta be defined as in equation 2.3, and that the discount rate could be determined as in equation 2.4. He also identified strategic issues such as flexibility and quality improvements, when to invest and when to abandon, and the value of learning to be considered subjectively as part of the final decision.

$$\beta_p = \beta_r\ [1 + (PV_{fc} / PV_p)] \tag{2.3}$$

where:

β_p : the beta of the project,
β_r : the beta of revenues associated with the proposed project,
PV_{fc}: the present value of fixed costs associated with the proposed project,
PV_p : the present value of assets associated with the project.

$$r_p = r_f + \beta_p \tag{2.4}$$

where:

r_p: risky discount rate,
r_f: risk free discount rate.

Kaplan (1986) suggested that DCF analysis might be used to evaluate CIM investments without considering the difficult-to-measure benefits in the analysis. If the result of the DCF analysis is positive, the investment is justified. If NPV is negative, then the difficult-to-measure benefits should be compared with the additional cash flows required to make the NPV positive. If management is willing to pay, at least the additional cash flows to enjoy the difficult-to-measure benefits, then the investment is justified. Naik and Chakravarty (1992) felt that Kaplan's suggestion could lead to inconsistency and inaccuracy. Also, this approach does not help in selecting from alternatives that satisfy the DCF criterion.

Bennett and Hendricks (1987) refined Kaplan's approach, suggesting that benefits not included in the first DCF analysis could be quantified using a team approach and probability analysis. Then, the viability of the project could be assessed using the standard cash flow methods.

Hundy and Hamblin (1988) advocated the IRR method because it gives a single value for project worth regardless of interest rate changes. They claimed that specific project risks can be easily modelled using spreadsheet methods and illustrated their approach using a numerical example based on a conventional machine shop and a flexible machining system as alternatives. Based on pessimistic, optimistic and most likely scenarios, they concluded that the FMS facility should be installed.

Although the use of DCF analysis has many advantages including low cost (Azzone et al., 1992), simplicity, clarity, and ease of data collection (Meredith and Suresh, 1986), there are many problems in applying it to the evaluation of investment decisions in AMT. Canada and Sullivan (1990) summarised these problems under three headings: arbitrary hurdle rates, comparison with the status quo, and insufficient benefit analysis.

Canada and Sullivan first noted that firms often use an after tax hurdle rate of 20% or more, thus penalising investments in AMT which tend to be more beneficial in the long-term. The second problem in applying DCF methods in evaluating AMT investment decisions is comparison with the status quo. This means that AMT investments are being implicitly compared to 'doing nothing' where 'doing nothing' presumes that the existing process will continue to perform as it has in the past. However, this is an unlikely scenario, a 'do nothing' firm is likely to lose market share. The third problem, insufficient benefit analysis, arises because of the difficulty of quantifying the qualitative benefits of AMT investments in cash flow terms. This problem was discussed in the previous chapter.

Mathematical Models

Miltenburg and Krinsky (1987) suggested a mathematical model to evaluate FMS utilising the DCF framework. Their suggestion was based on the application of an iterative 'decision analysis cycle' (figure 2.3). The procedure is divided into three phases. In the first (deterministic) phase, the variables affecting the decision are defined, relationships between them developed, and the importance of each variable determined. In the second (probabilistic) phase uncertainty is analysed by specifying probabilities for the more important variables. Then the associated probability distribution of the outcome variables is determined. The final (informational) phase uses the results of the first two phases to determine the economic value of eliminating uncertainty in each of the important variables. If there are profitable sources of information, new information should be gathered and the original analysis revised. The analysis is repeated until there are no longer any profitable sources of new information. Then, from the probability distributions of the outcome variable for each alternative, the principles of stochastic dominance and risk preference are used to determine the best decision.

The most appealing feature of this model is the formal incorporation of a method for updating the state of knowledge by searching iteratively to determine usefulness of new information (Mensah and Miranti, 1989, p. 196). On the other hand, only financial variables are considered in the NPV analysis.

Park and Son (1988) developed a multi-period net present value linear programming model. They proposed measurements for the costs of quality

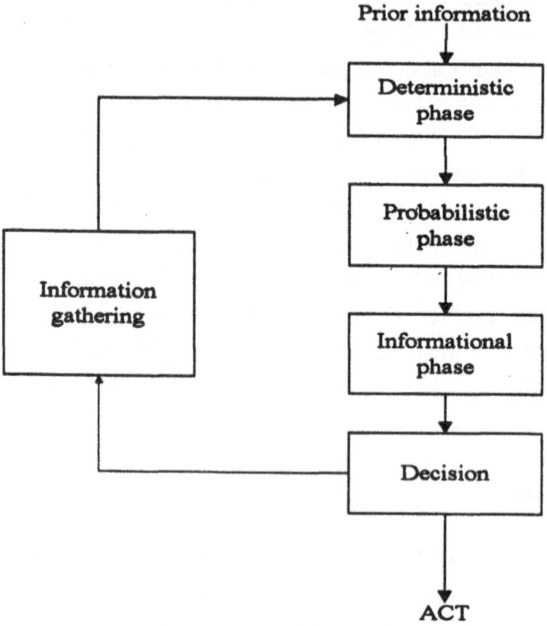

Figure 2.3 The decision analysis

Source: Adapted from Miltenburg and Krinsky, 1987.

(prevention, appraisal, and failure), and flexibility (idle, set-up, waiting, and inventory). They also explicitly considered opportunity costs in the decision model. They distinguished between capital performance (NPV1) which refers to the net present value of the initial investment, its depreciation and net salvage value; and production performance (NPV2) which refers to the net present value of benefits resulting from manufacturing and marketing. They formulated a linear programming model that maximises NPV1 + NPV2 (the objective function) subject to various constraints representing the limited resources and the logistical relationship between the manufacturing and inventory systems. The model is illustrated through a case study comparing a batch shop with a proposed cellular manufacturing system.

An advantage of linear programming is the ability to consider many constraints, to calculate an optimal solution and provide sensitivity analysis. However, the model is restricted by the assumptions of linear programming, especially the assumed linear relationship between variables. Also, like other quantitative models, precise estimates for all the variables are required; and, non-quantifiable factors cannot be considered in the model. On a technical point, there is duplication in Park and Son's analysis because depreciation is discounted as if it is part of cash flow while its total value is also taken into account in the initial investment.

The Qualitative Approach

Some approaches to AMT decision making concentrate mainly on qualitative variables. The approaches can be divided into:

1. Strategic Analysis.
2. Differentiated Gradualist Approach.

Strategic Analysis

Meredith and Suresh (1986) described a strategic approach based on four criteria: technical importance, business objectives, competitive advantage, and research and development (R&D). Justification of projects is related to the four criteria. Some projects might be approved in order to achieve desirable technical ends. Similarly, projects may be executed if they are essential to achieve business objectives. For example, a project may be approved in order to give a good impression to customers, vendors, and visitors. Similarly, to gain competitive advantage, the firm may invest in an automation project. Finally, under the R&D criterion, the firm approves R&D investments if they hold sufficient strategic promise. It is presumed that, as there are many such projects, some will eventually come through to provide returns which reimburse the failures.

Medearis et al. (1990) introduced an alternative strategic framework. Based on Porter's (1980) model, they argued that the forces that drive industry competition include the threat of new entrants, the bargaining power of suppliers and customers, the threat of substitute goods, and the rivalry existing amongst industry participants. They argued that the profitability of a company is dependent on the strongest of these forces and the dominant force must be considered in developing a competitive strategy. They analysed the performance of FMS in terms of Porter's five forces and concluded that FMS offers organisations an opportunity to improve competitive positioning within an industry and even to expand the definition of the industry.

Perhaps the most holistic attempt to develop a model for selecting appropriate FMS using a strategic approach was the study by Elango and Meinhart (1994). They argued that the selection and implementation of an FMS should be consistent with the strategy and environment of the firm. The analysis begins by identifying industry and market needs using Porter's model. The strength of the five competitive forces determines the ability of the firm to earn above average profits. The manufacturing strategy that should be developed should take account of the FMS strategies pursued by the dominant or majority of firms in the industry. They defined 'manufacturing strategy' as the strategic capability offered by the manufacturing operations of the firm in order to facilitate the implementation of overall corporate and business strategy. The next step is the analysis of strategic flexibility to determine the required amount of flexibility that should be evaluated with respect to the financial performance of this system[3]. They advocated the use of the standard financial techniques such as return on investment, NPV, IRR, or value-based approaches for the financial evaluation of the system. The final stage of the decision analysis for selecting an FMS is to determine whether the capabilities of the organisation match the skills required to run the system effectively.

Differentiated Gradualist Approach

Mensah and Miranti (1989) in their review of capital expenditure analysis and automated manufacturing systems described what they called a 'differentiated gradualist approach' for the evaluation of automation projects. According to this approach, the first project is regarded as a learning experience and not subjected to rigorous DCF analysis. The evaluation of subsequent projects is modified on the basis of the better understanding of cost patterns apparent through the review of the initial project's outcome. Then, with greater experience it is expected that the DCF framework can be further refined.

They claimed that the merits of this approach are twofold. First, it permits a better perspective to be developed for evaluating the qualitative benefits of automation. Second, since many firms implementing automated systems undergo an extensive trial-and-error period, the pilot approach can provide valuable information which will allow costly errors in full-scale manufacturing to be avoided. Mensah and Miranti noted that this is a costly approach which may be appropriate only for large firms with substantial resources.

Although the qualitative approaches address the major problem in applying traditional investment models to AMT, they depend mainly on subjective judgements and are thus not very 'scientific'. They also normally ignore financial analysis and it therefore seems unlikely that they can be used as the sole basis for AMT investment decision making.

The Integrated Approach

These methods for evaluating investment decisions in AMT take into account a wider range of variables, both qualitative and quantitative, in the analysis. Two groups can again be distinguished:

1. Scoring models.
2. Analytical hierarchy process.

Scoring Models

Scoring models are a set of techniques which reduce a multi-attribute problem to a single dimension (Randhawa and West, 1992). There are three basic techniques: unweighted 0-1 scoring, unweighted factor scoring and weighted factor scoring.

Meredith and Suresh (1986) described the unweighted 0-1 scoring model. After selecting the relevant criteria, each alternative is rated 0 or 1 with respect to each criterion. The summation of these rates for each alternative is used to rank alternative projects. See the example in table 2.1. Simplicity is the main advantage of this technique. However, it has two disadvantages. First, it assumes all criteria are of equal importance. Second, it does not allow for gradation of the degree to which a specific alternative meets the various criteria.

Nelson (1986) suggested an unweighted factor scoring model for evaluating and prioritising AMT projects. This model encompasses the scores for five variables: technology assessment (t), equipment evaluation (e), capacity elasticity (c), cost-budget analysis (b), and adjusted net present value (v). The total score of a project (s) can be obtained from the linear combination of these variables calculated as relative scores (denoted *) as shown in equation 2.5 [4].

$$s_i = t_i^* + e_i^* + c_i^* + b_i^* + v_i^* \qquad (2.5)$$

Table 2.1 An example for unweighted 0-1 factor model

	Alternatives		
Factors	**A**	**B**	**C**
ROI 25%	1	1	0
Compatibility	1	1	1
Quality improvement	0	1	1
Volume flexibility	0	0	1
Product flexibility	0	0	1
Total score	2	3	4*

* The best alternative.

where:

s_i : the score for project i, $0 < s_i \leq 5$,
t_i^* : the technology assessment score for project i,
e_i^* : the equipment evaluation of shop i,
c_i^* : the relative workload elasticity of capacity for shop i,
b_i^* : the relative cost-to-budget ratio score for project i,
v_i^* : the relative adjusted net present value score for project i.

The technology assessment score (t_i) is a linear combination of three variables: emphasis, status, and impact. The emphasis (t_{i1}) measures the concept of strategic need from a local facility view, and depends for its score value upon the current availability, adequacy and quality of the proposed technology. The status (t_{i2}) measures the state of knowledge of the proposed technology and its use in the market place. The impact (t_{i3}) measures the benefit directly attributable to the proposed project. The summation of these three scores produces a total technology assessment score as in equation 2.6. The relative score (t_i^*) is the quotient of t_i divided by the maximum t_i achieved from among the i projects (t_m) as in equation 2.6a.

$$t_i = t_{i1} + t_{i2} + t_{i3} \tag{2.6}$$

$$t_i^* = \frac{t_i}{t_m} \tag{2.6a}$$

Similarly, the equipment evaluation score (e_i) is a linear combination of three variables: condition, age, and suitability. The condition (e_{i1}) is a weighted average condition score for all items of equipment in shop i. The condition score for each item is the evaluator's estimate of the cost of repair of the equipment as a percent of the total replacement cost of the item. The age (e_{i2}) is a weighted average score for all items of equipment in shop i. The suitability (e_{i3}) is a weighted average sum of two 0-1 parameters. These parameter scores measure the general availability of proper equipment for relevant shop functions. The summation of the three scores produces a total equipment evaluation score as in equation 2.7. The relative score (e_i*) is the quotient of e_i divided by the maximum e_i achieved from among the i shops (e_m) as in equation 2.7a.

$$e_i = e_{i1} + e_{i2} + e_{i3} \tag{2.7}$$

$$e_i^* = \frac{e_i}{e_m} \tag{2.7a}$$

The elasticity of capacity measures sensitivity of planned changes in capacity. To produce the capacity elasticity score (c_i*), the absolute value of the difference of the elasticity value from 1 for each shop i is transformed by dividing it by the minimum absolute value, minus 1, from amongst the i shops, the inverse of this quotient is c_i* as in equation 2.8.

$$c_i^* = \frac{\min |c_i - 1|}{|c_i - 1|} \tag{2.8}$$

where:

c_i^* : the relative elasticity score for shop i, $0 < c_i^* \leq 1.0$,

$|c_i - 1|$: the absolute value of the difference of c_i from 1 for shop i,

$\min |c_i - 1|$: the minimum absolute value from amongst the i shops.

The cost-budget ratio (b_i) measures the degree to which the total estimated project investment is within the limit of estimated funding. To calculate the cost-budget ratio score (b_i*), cost-budget ratios b_i are transformed by dividing by the minimum cost-budget ratio from among the i projects (b_{min}), and inverting this quotient as in equation 2.9.

$$b_i^* = \frac{b_{\min}}{b_i} \tag{2.9}$$

To calculate the NPV score (v_i*), first the net present value is adjusted to consider the risk and interdependence of the projects. In the case of risky independent projects, the cost of capital (k_i) is adjusted to vary directly with risk, and NPV is weighted by an explicit risk factor (p_i). In the case of independent projects, p_i is the simple probability of technical success, but in the case of interdependent projects, the probability of success is conditional upon the probability of funding or success of dependent projects (Nelson, 1986). Then, the adjusted NPV (v_i) is transformed by dividing v_i by the maximum adjusted NPV (v_m) from among the i projects as in equation 2.10.

$$v_i^* = \frac{v_i}{v_m} \tag{2.10}$$

The main advantages of Nelson's scoring model are its treatment of non-economic criteria and project interdependency. On the other hand, the disadvantages of this model are twofold. First, it considers all variables to have equal weight. Second, there is duplication in two terms since the capital cost of the project is considered in the calculation of the cost-budget ratio score and the NPV score.

Parsaei and Wilhelm (1989) proposed a weighted factor scoring model to evaluate AMT alternatives based on a linear additive model developed by Klee (1971). The model involves two stages. In the first stage, the desirability of various strategic (long-term) automation proposals is examined. In the second stage, each tactical (short-term) alternative for implementing the selected strategic proposal is evaluated. They proposed two sets of weighting scales, one for ranking the importance of strategic and tactical alternatives, and the other for evaluating project performance against each strategic and tactical alternative. These scales are then employed in a linear additive model at each stage. An illustrative example is presented in table 2.2.

This model overcomes one of the disadvantages of Nelson's scoring model; factors are weighted according to their importance. However, on the other hand, some financial factors are treated as intangibles and many other financial factors are overlooked.

In general, scoring models are attractive for three reasons. First, they can incorporate management's value judgements regarding intangible factors. Second, they can reflect management's changing priorities and objectives in the evaluation process through changes in the weights assigned to various factors. Third, they permit the application of mathematical programming techniques to identify the most desirable projects in relation to specific decision criteria (Mensah and Miranti, 1989). However, the proposed models assume independent criteria and linear relationships and are sensitive both to the variables identified and to the relative weights subjectively attached to them. Furthermore, there is no means of measuring the consistency of decision makers' judgements. Some models might also be criticised because of duplication of criteria. For example, in Nelson's model, the project's capital cost appeared in both the cost-budget analysis and in the adjusted NPV.

Analytic Hierarchy Process (AHP)

The 'analytic hierarchy process', whilst retaining all of the advantages of simpler scoring methods, allows considerably more sophisticated decision models to be constructed and has therefore been suggested as an appropriate method of combining the multiple attributes of AMT projects. The Analytic Hierarchy Process (AHP) was originated during the 1970s, and its theory is now well developed (see, for example, Saaty, 1990). The AHP enables decision makers to model complex decisions such as the decision to invest in AMT, allowing concurrent interactions between variables to be structured in a hierarchical format. It is based on the three stages of problem solving: decomposition, comparative judgements, and synthesis of priorities. The inputs to the AHP are judgements about the relative importance of each criterion and the evaluation of each decision alternative against each criterion. Analysis proceeds from the lowest level of the hierarchy, and determining the importance of low level effects and judgements on the overall goal of the hierarchy. The output from the process is a ranking

Table 2.2 Illustrative example for Parsaei and Wilhelm's model

Step 1. Selection of the best strategic automation proposal

i	Weigh W_i	Attribute	Proposal 1 X_{i1}	Proposal 1 W_iX_{i1}	Proposal 2 X_{i2}	Proposal 2 W_iX_{i2}	Proposal 3 X_{i3}	Proposal 3 W_iX_i
1	0.09	Reduced lead times	0.60	0.05	0.60	0.05	0.80	0.07
2	0.14	Reduced WIP	0.60	0.08	0.80	0.11	0.80	0.11
3	0.18	Improved quality	0.80	0.14	0.60	0.11	1.00	0.18
4	0.18	Cost reduction	0.80	0.14	1.00	0.18	1.00	0.18
5	0.14	Product flexibility	0.40	0.06	0.60	0.08	0.80	0.11
6	0.14	Schedule flexibility	0.60	0.08	0.40	0.06	0.20	0.03
7	0.14	Safety important	0.40	0.06	0.60	0.08	0.80	0.11
		P_j		0.69		0.67		0.79[a]
		Normalised P_j		0.87		0.85		1.00

Step 2: Selection of the best tactical alternative

i	Weight W_i'	Attribute	Alternative 1 X_{i1}'	Alternative 1 $W_i'X_{i1}$	Alternative 2 X_{i2}'	Alternative 2 $W_i'X_{i2}$	Alternative 3 X_{i3}'	Alternative 3 $W_i'X$
1	0.21	Annual cost recovery	1.00	0.21	0.80	0.17	0.60	0.13
2	0.16	Software flexibility	0.60	0.10	0.60	0.10	1.00	0.16
3	0.05	Required floor space	0.60	0.03	0.80	0.04	0.80	0.04
4	0.16	Ease of operation	0.80	0.13	0.60	0.10	0.60	0.10
5	0.11	Manpower required	0.80	0.09	0.60	0.07	0.40	0.04
6	0.16	Compatibility	0.60	0.10	0.80	0.13	1.00	0.16
7	0.16	Future expansion	0.40	0.06	0.80	0.13	1.00	0.16
		A_j'		0.72		0.74		0.79[b]
		Normalised A_j'		0.91		0.94		1.00

[a] the most desirable strategic proposal ($Max\ P_j = \Sigma W_i X_{ij}$)

[b] the best available alternative ($Max\ A_j' = \Sigma W_i' X_{ij}'$)

Source: Adapted from Parsaei and Wilhelm, 1989.

denoting the overall preference for each of the decision alternatives (Anderson et al., 1994).

The analytic hierarchy process has been widely recommended for AMT decision making and is a key element of the algorithm developed in later chapters. It is, therefore, important to understand AHP principles and its use

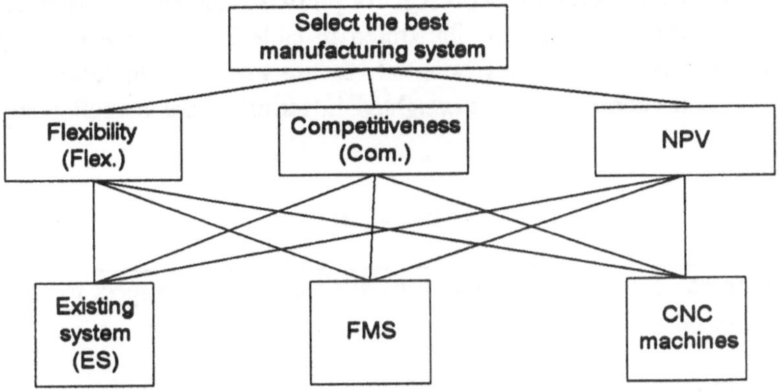

Figure 2.4 Developing the problem hierarchy

is illustrated in the following example. The first step is to develop a diagrammatic representation of the problem in terms of the overall goal or focus, criteria, and the decision alternatives. This diagram represents the hierarchy for the problem as, for example, in figure 2.4.

The second step requires pairwise comparisons using the AHP measurement scale (table 2.3) to establish priority measures for both the

Table 2.3 The measurement scale of AHP

Numerical value	Definition
1	Equal importance of both elements.
3	Moderate importance of one element over another.
5	Strong importance of one element over another.
7	Very strong importance of one element over another.
9	Extreme importance of one element over another.
2,4,6,8	Intermediate values between two nearby judgements.
Reciprocals	Used to reflect dominance of the second element as compared with the first.

Source: Harker, 1989.

criteria and decision alternatives. A set of pairwise comparison matrices may be developed at this stage, as illustrated in table 2.4. For the hierarchy of figure 2.4, the three decision alternatives are evaluated in terms of each criterion and each criterion is weighted in terms of the overall goal. The

Table 2.4 The results of pairwise comparisons

A. Pairwise comparisons of the three alternatives with respect to each criterion

Flexibility

	ES	**FMS**	**CNC**	**Priority Vector**	
ES	1	1/7	1/3	0.083	
FMS	7	1	5	0.723	
CNC	3	1/5	1	0.194	
					CR=0.058

Competitiveness

ES	1	1/4	1/3	0.118	
FMS	5	1	2	0.576	
CNC	3	1/2	1	0.306	CR=0.078

NPV

ES	1	7	4	0.701	
FMS	1/7	1	1/3	0.086	
CNC	1/4	3	1	0.213	
					CR=0.028

B. Pairwise comparisons of the criteria with respect to overall goal

	Flex.	**Com.**	**NPV**	**Priority Vector**	
Flex.	1	1/3	1/2	0.156	
Com.	3	1	4	0.619	
NPV	2	1/4	1	0.225	
					CR= 0.095

priority vector is computed as an average of the elements in each row of the normalised pairwise comparison matrix[5].

The weight assigned to each decision alternative and criterion may be estimated from data or subjectively by decision makers. In the latter case, it would be desirable to measure the consistency of decision makers' judgements and the AHP provides such a measure through the consistency ratio (*CR*) which is an indicator of the reliability of the model (Apostolou and Hassell, 1993). This ratio is designed in such a way that values of the ratio exceeding 0.10 indicate inconsistent judgements; and, in such cases, it is probably preferable to revise the original values in the pairwise comparison matrix (Anderson et al., 1994). To illustrate the computation of *CR*, let the pairwise comparison matrix be denoted X_1 and the priority vector be denoted X_2. First calculate $X_3 = X_1 * X_2$. For the pairwise comparison matrix of the three alternatives with respect to flexibility:

$$X_3 = \underline{\text{Pairwise comparison matrix}} \times \underline{\text{Priority vector}}$$

$$\begin{matrix} \text{ES} \\ \text{FMS} \\ \text{CNC} \end{matrix} \begin{bmatrix} 1 & 1/7 & 1/3 \\ 7 & 1 & 5 \\ 3 & 1/5 & 1 \end{bmatrix} \times \begin{bmatrix} 0.083 \\ 0.723 \\ 0.194 \end{bmatrix}$$

$$\begin{aligned} X_3 = \; & 1 \times 0.083 + 1/7 \times 0.723 + 1/3 \times 0.194 = 0.251 \\ & 7 \times 0.083 + 1 \times 0.723 + 5 \times 0.194 = 2.274 \\ & 3 \times 0.083 + 1/5 \times 0.723 + 1 \times 0.194 = 0.588 \end{aligned}$$

Then, calculate each element of X_4 by dividing each element of X_3 by the respective element in the priority vector X_2:

$$\begin{aligned} X_4 = \; & 0.251 \div 0.083 = 3.024 \\ & 2.274 \div 0.723 = 3.145 \\ & 0.588 \div 0.194 = 3.031 \end{aligned}$$

Next, the average of the elements of X_4 can be denoted as λ_{max}:

$$\lambda_{max} = \frac{3.024 + 3.145 + 3.031}{3} = 3.067$$

Then, the consistency index *CI* is determined by:

$$CI = \frac{\lambda_{max} - n}{n-1} = \frac{3.067 - 3}{2} = 0.0335$$

where: *n* is the number of the elements of priority vector. Finally, the *CR* is determined by:

$$CR = \frac{CI}{RI} = \frac{0.0335}{0.58} = 0.058$$

where: *RI* is a random index computed for matrices of order $n = 1$ to 15 via simulation (Apostolou and Hassell, 1993). Results of this simulation are included in table 2.5.

In this example the CR is less than 0.10, and so it is concluded that the judgements are sufficiently consistent with each other.

The final step in the AHP is to develop an overall priority ranking. This can be obtained by summing the products of the criterion priority and the priority of its decision alternative. For example:

<u>Priority of the alternatives matrix</u> <u>Criterion priority</u> <u>Overall priority</u>

$$\begin{matrix} ES \\ FMS \\ CNC \end{matrix} \begin{bmatrix} 0.083 & 0.118 & 0.701 \\ 0.723 & 0.576 & 0.086 \\ 0.194 & 0.306 & 0.213 \end{bmatrix} \times \begin{matrix} Flex \\ Com \\ FP \end{matrix} \begin{bmatrix} 0.156 \\ 0.619 \\ 0.225 \end{bmatrix} = \begin{matrix} ES \\ FMS \\ CNC \end{matrix} \begin{bmatrix} 0.243 \\ 0.489 \\ 0.268 \end{bmatrix}$$

After obtaining the overall priority of the decision alternatives, the decision maker can choose the alternative which has the highest priority. In this example, the decision maker should choose the FMS alternative as it has the highest overall priority (0.489).

The ability of the AHP to structure a complex problem hierarchically and synthesise priorities for decision alternatives, has attracted many

Table 2.5 Random inconsistency index (*RI*)

n	1	2	3	4	5	6	7	8	9	10	11	12	13	14	15
RI	0.00	0.00	0.58	0.90	1.12	1.24	1.32	1.41	1.45	1.49	1.51	1.48	1.56	1.57	1.59

Source: Golden et al., 1989, p. 32.

researchers interested in modelling AMT investment decisions. Several studies are reviewed in the following sections.

Wabalickis's Four Level Hierarchy Wabalickis (1988) considered a hierarchy of four levels: overall goal, criteria, subcriteria and decision alternatives. The overall goal is to select the best manufacturing system from three alternatives: existing system, FMS and NC machines. The criteria and subcriteria are:

Criteria	Sub-criteria
Inventory costs	reduced lead time; minimize inventory
Sales	product mix response; profitability; product change response
Operating costs	tooling ; scrap and start-up costs

Wabalickis introduces this hierarchy as an example. Nevertheless it is typical of many of the proposals in the literature and suffers from a number of defects. First, there is no obvious reason why the particular criteria of inventory costs, sales and operating costs have been chosen as important decision making criteria. Second, if costs are important, it is hardly necessary to rely on subjective pairwise comparisons in order to analyse them; traditional economic methods would appear to be more suitable. Third, there is duplication in the hierarchy because profitability is partly dependent on both inventory costs and operating costs. Duplication is likely to lead decision makers to overweight certain criteria, in this case, financial measures. Given that Wabalickis only claims this example as illustrative, these comments may seem unduly critical. However, it will be seen that similar criticisms can be levelled at other models reported in the literature.

Srinivasan and Millen's Four Level Hierarchy Srinivasan and Millen (1986) used the AHP to weight qualitative variables not reflected in cash flow estimates. Decision alternatives were then ranked according to their relative strategic net present value (*SNPV*) computed as in equation 2.11. The decision alternative with the highest *SNPV* is recommended.

$$SNPV_i = (PVCF_i \times W_i) - NINV_i \qquad (2.11)$$

where:

$SNPV_i$: relative strategic net present value for decision alternative i,
$PVCF_i$: present value of the estimated cash flows for decision alternative i,
W_i : the relative strategic weight for decision alternative i (from the AHP),
$NINV_i$: net capital investment for decision alternative i.

Srinivasan and Millen furnished an example to illustrate their model. They assumed that three technological alternatives were to be considered: the existing system, FMS 1, and FMS 2. A hierarchy representing the qualitative factors is shown in figure 2.5. Using their assumptions, Srinivasan and Millen derived final weights of 0.03, 0.68, and 0.29 for the three alternatives respectively. The *PVCFs* for the three alternatives were \$2.0, \$3.2, and \$3.3 million and *NINVs* for the three alternatives were \$0.1, \$2.0, and \$1.5 million. Applying the equation 2.11, the strategic net present value (*SNPV*) for the three alternatives can be calculated:

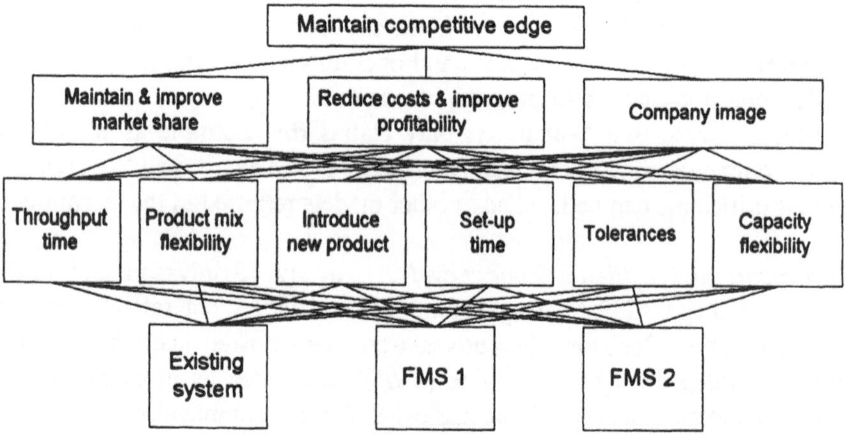

Figure 2.5 Srinivasan and Millen's hierarchy example

$SNPV_{\text{existing system}} = (2.0 \times 0.03) - 0.1 = -\$0.04\text{m}$
$SNPV_{\text{FMS 1}} = (3.2 \times 0.68) - 2.0 = +\$0.1176\text{m}$
$SNPV_{\text{FMS 2}} = (3.3 \times 0.29) - 1.5 = -\$0.543\text{m}$

∴ FMS 1 is recommended.

It should be noted that the calculations of *SNPV* follow the suggestion of Srinivasan and Millen (1986, p. 88). They actually calculated *SNPV* in a different way (p. 89):

$SNPV_{\text{existing system}} = (2.0 - 0.1) \times 0.03 = \$0.057\text{m}$
$SNPV_{\text{FMS 1}} = (3.2 - 2.0) \times 0.68 = \$0.816\text{m}$
$SNPV_{\text{FMS 2}} = (3.3 - 1.5) \times 0.29 = \$0.522\text{m}$

In this particular example the same conclusion is reached, but this would not necessarily always be so.

As calculated by Srinivasan and Millen, the weights of the qualitative benefits affect both the *PVCF* and *NINV*. This is surely an error because the initial investment is known with some certainty and multiplying it by any weighting factor must be spurious. It is also difficult to see the justification for adjusting *PVCF* by the weighting factors as initially recommended. Also, there is duplication in the calculation of *SNPV* because the qualitative factors in the hierarchy include some financial factors (such as 'reduce cost & improve profitability') which should only be reflected in the calculations of NPV.

Chandra and Schall's Mathematical Programming Model Chandra and Schall (1988) used the Leontief input-output model and linear programming to determine the input of an AHP for evaluating two alternative flexible manufacturing systems. The hierarchy had three levels; the first level represents the decision: choose the design with the minimum cost, the second level represents the attributes: capital cost, total sector multipliers[6], expected maximum number of finished items, and expected maximum number of scrapped items, and the last level represents the two alternatives: design I and design II.

In the light of the output of both the Leontief model and the linear programming model, in addition to the judgement of the decision makers, the priorities of the two alternatives are determined. Since the focus of the hierarchy is to minimise the total cost, the design with the lowest priority should be recommended.

An advantage of this approach is that it provides the AHP with more precise inputs generated by the Leontief and linear programming models. However, the approach is limited by the linearity assumptions of both the Leontief and the linear programming models. A second drawback of the approach is its emphasis on cost factors, so it does not address most of the key issues in AMT decision making.

Putrus's Three Dimensions Approach Putrus (1990) treated the justification problem as a three dimensional framework of risk, benefits (tangible and intangible) and time (short-, medium-, and long-term). He aimed to justify a CIM project based on the ratio of total benefits to total risk for different time intervals. In order to calculate the total benefits to total risk ratio a company hierarchy was constructed (see figure 2.6) for the

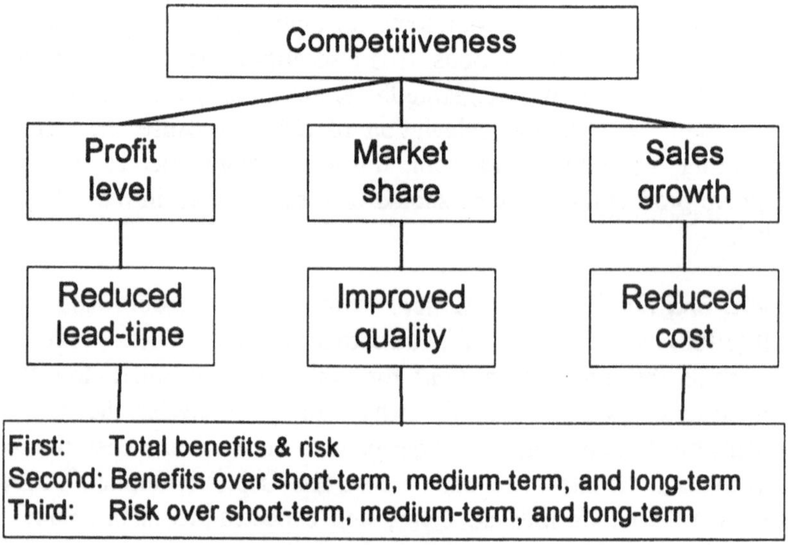

Figure 2.6 Putrus's model hierarchy

Source: Adapted from Putrus, 1990.

project under consideration. The lowest level of this hierarchy depends on which step of the calculations is being considered.

First, the lowest level is defined as total benefits and risk as if they are two alternatives. Then the final ranking of total benefits and risk can be calculated by normal AHP calculations.

Second, the lowest level is redefined as three alternatives of benefits over short-term, medium-term and long-term. Thus, the final ranking of total benefits over these three periods of time can be calculated.

Third, the lowest level is redefined as three alternatives of risk over short-term, medium-term and long-term, and the final ranking of risk over these three periods of time can be calculated.

After these three definitions of the lowest level of the hierarchy, there will be three groups of values: two final rankings for total benefits and risk, three final rankings for total benefits in the short-term, medium-term and long-term, and three final rankings for total risk in the short-term, medium-term and long-term. The next step is to calculate a ratio of total benefits to total risk values by dividing the final rankings of total benefits by total risk calculated in the first definition of the lowest level of the hierarchy. Then, this ratio is used to adjust the three final rankings of risk: short-term, medium-term and long-term (from the second definition of the lowest level of the hierarchy) by dividing each value of risk by this ratio. This produces three adjusted risk values. Finally, the three final rankings of total benefits: short-term, medium-term and long-term (from the third definition of the lowest level of hierarchy) are divided by the three adjusted risk values calculated in the previous step. This calculation produces three different ratios of total benefits to total risk, each representing different time intervals. A ratio greater than one means the benefits of the CIM project outweigh the risk of CIM for that term. A ratio less than one means the risk outweighs the benefits for that term. A numerical example to illustrate the calculations of total benefits to total risk ratio is provided in table 2.6.

There are two main criticisms of the calculation of the ratio of total benefits to total risk. The first criticism relates to the construction of the project hierarchy. There is no obvious relation between the high level criteria: profit level, market share, sales growth and the suggested sub-analysis: reduced lead-time, improved quality and reduced cost. And there are a number of inter-relationships between criteria. Profit and cost are obviously related and so, arguably, are profit, market share and sales growth. The second criticism relates to the method of calculating the total benefit/total risk ratio - for which there is no theoretical justification. For

Table 2.6 A numerical example for Putrus's model

Assume a company evaluates a flexible manufacturing cell project. The starting point is to construct a hierarchy as in figure (2.6). At each step, the lowest level of the hierarchy depends on the calculation of total benefits to total risk ratio.

1. Let the lowest level of the hierarchy be two alternatives of total benefits and risk. Suppose after constructing pair-wise comparisons for each level and using the AHP calculations, the final rankings of total benefits and risk are 0.33 and 0.67.
2. Let the lowest level of the hierarchy be three alternatives of total benefits in the short-term, medium-term, and long-term; and suppose the final rankings are 0.22, 0.33, and 0.45.
3. Let the lowest level of the hierarchy be three alternatives of total risk in the short-term, medium-term, and long-term; and suppose the final rankings are 0.052, 0.322, and 0.626.
4. Ratio of total benefits to total risk = 0.33 ÷ 0.67 = 0.493.
5. Adjusted risk values:
   short-term = 0.052 ÷ 0.493 = 0.105
   medium-term = 0.322 ÷ 0.493 = 0.653
   long-term = 0.626 ÷ 0.493 = 1.270
6. Total benefits to total risk ratio:
   short-term = 0.22 ÷ 0.105 = 2.095
   medium-term = 0.33 ÷ 0.653 = 0.505
   long-term = 0.45 ÷ 1.270 = 0.354

∴ The project's benefits outweigh its risk in the short-term period while they do not outweigh its risk either in the medium-term period or in the long-term period.

example, the meaning of 'adjusted risk values' as calculated (by dividing the three final rankings of total risk in the second definition of the hierarchy by the ratio of total benefits to total risks in the first definition of the hierarchy) is unclear.

Ghosh and Wabalickis's Comprehensive, Comparative Analysis Ghosh and Wabalickis (1991) introduced an approach that they called comprehensive, comparative analysis. They suggested that two separate hierarchies be constructed; for costs and benefits of the manufacturing systems respectively. In each hierarchy they identified four major elements: workforce, equipment, methods/materials and information processing, and a further sub-level defined the manner in which these factors were to be measured. At the lowest level were the decision alternatives. Analysis of both the cost and the benefit hierarchies allowed a cost/benefit ratio to be constructed for each alternative, the minimum value of the cost/benefit ratio giving the optimum manufacturing system.

In this model Ghosh and Wabalickis claim the four elements as 'the basic components of any system' and go on to identify their characteristics in considerable detail. For example, the project's impact on the workforce is to be analysed in terms of age, job fit, skills, number, expense and motivation. Whilst analysing production systems in detail, Ghosh and Wabalickis are dismissive of more 'strategic' concerns:

> '... The basic benefits of the production system are more realistically estimated than strategic factors such as financial position, competitive position or sales increases. Any attempt to place a score on these items is ill-advised and mere guesswork'. (p. 5)

This may contain an element of truth, but this has not deterred many other researchers! In passing it is noted that Ghosh and Wabalickis furnish an example where some of the costs identified would be excluded from conventional financial analysis because they are 'sunk'. Also, some of the example weightings seem highly unlikely. They assume, for example that 'management' rates 'workforce considerations' as four times more important than 'equipment' or 'materials' issues and five times as important as 'information systems'.

Naik and Chakravarty's Three Level Hierarchy Naik and Chakravarty (1992) suggested a framework for evaluating AMT investments based on three hierarchical levels of evaluation. First, a strategic level of evaluation which aims to select the best category of manufacturing system consisting of appropriate system modules at desirable levels of integration in order to provide desired manufacturing capabilities. This level reduces the wide range of available options. Second, an operational level of evaluation which aims to identify the top two or three alternatives meeting the option selected in the first level. This identification would be based on the received vendor quotes. The third level of evaluation is the financial evaluation level in which detailed discounted cash flow analysis is conducted for the systems chosen at the operational level. The system with the best financial performance is selected if it meets the performance requirements of the firm. Otherwise, the evaluation process must be repeated from the strategic level.

Datta et al.'s Empirical Study Datta et al. (1992) developed an analysis based on a study of the literature and of consultations with managers at a case study site. They based their analysis on a questionnaire completed by a sample of fifteen managers and identified 12 factors to be weighted in the decision. The factors are reported in descending importance in table (2.7). This empirically grounded analysis is more convincing but appears to suffer from failure to include any financial analysis. This might not be disturbing if all financial factors had been excluded so that they might be analysed separately, but the identified factors include two financial variables: 'potential for reduction in inventory costs' and 'investment'.

O'Brien and Smith's Strategic Model O'Brien and Smith (1993) developed a strategic modelling tool which relies on the AHP to evaluate the investment in new hardware and software technologies in electronics industries. They

Table 2.7 Weight of factors considered in AMT projects

Factor	Weighting
Quality and reliability	0.21
Flexibility	0.20
Throughput	0.13
Capacity utilization	0.12
Inventory costs	0.11
Investment	0.07
Market position	0.04
Technology position	0.04
Technical feasibility	0.03
Employee relations	0.02
Human factors	0.02
Information generation	0.01
Total	1.00

Source: Adapted from Datta et al., 1992.

considered five strategies: business systems strategy, marketing strategy, servicing strategy, design and manufacturing strategy, and financial control strategy. The structure of their model is a hierarchy. At the top level the objective of the company is to select the best design system, at the intermediate level the company has to define the various strategic criteria which the best design system is expected to support, and to weigh the relative importance of these criteria. Finally, at the bottom level the model requires the user to compare and score two options (new technology and present technology) as regards the extent to which each option is expected to support the criteria. To assist the user in evaluating the extent to which each of the options supports the various strategies, they suggested subdividing the strategies into a number of sub-criteria, see figure 2.7.

The model is based on a two-stage process. First, the relative importance of the various criteria and sub-criteria is determined by carrying out a pairwise comparison of the criteria using the scale of 1-9. The score represents the degree of importance of one factor over the other. Second, based on the relative weighting for the various criteria, each of the investment options is compared and given a score for each of the criteria. The final output of this model is a score for each of the options which combines the assessment of its relative value for each

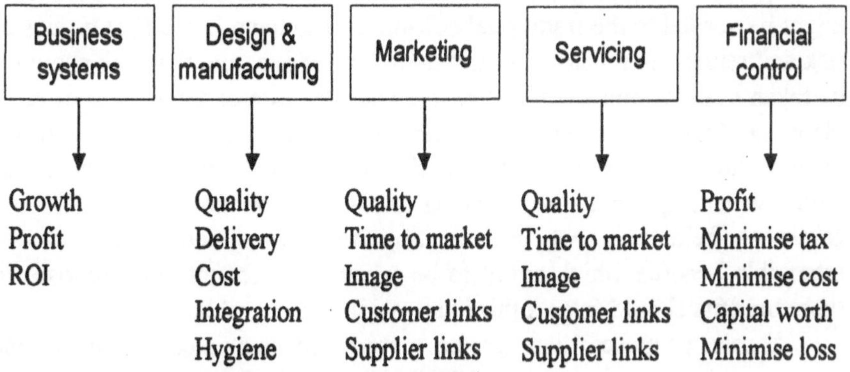

Figure 2.7 The sub-criteria for each main criterion (strategy) in O'Brien and Smith's model (1993)

criterion or sub-criterion for each of the strategies with the relative weighing of the associated strategies. Then, each strategy is represented by a separate hierarchy.

O'Brien and Smith claimed that this model can be used in evaluating AMT investments either as a substitute for financial appraisal or as a support for it whether before or after analysing the financial consequences of such investments. However, this claim needs to be carefully assessed because their suggested hierarchy suffers from the same problems as many others. There is duplication in the hierarchy with profit appearing twice together with various references to cost and ROI, which are not listed under 'financial control', as expected. In addition, the marketing and servicing criteria are duplicated and quality appears three times. It should be recognised that a given sub-criterion might have a different meaning when related to different decision dimensions; nevertheless, there must be considerable scope for confusion and duplication in a model such as this.

Accola's Analysis of Risk, Return and Intangible Benefits A most cogently argued analysis is provided by Accola (1994) who employed three major dimensions in developing a model for AMT decision making: financial risk, financial return and qualitative factors. Unlike other authors, Accola takes existing financial theory as his starting point and considers additions which might be needed to the traditional risk-return framework. He first argues that risk and return are intrinsically different dimensions, and that risk should not be taken into account by adjusting the discount rate or target rate of return. Here, he follows Ronen and Sorter (1972), who pointed out that such adjustments are likely to reflect the decision maker's *attitude* to risk rather than intrinsic project risk. Accola then goes on to reiterate the now commonly held view that, in evaluating AMT investments, there are often intangible benefits which ought to be taken into account. He proposes that these benefits should form a third dimension in the evaluation.

Accola proposes that project return be evaluated using a single measure: return on investment (ROI). Three sub-dimensions are proposed for the analysis of risk: variability of project outcomes (measured by standard deviation), possibility of ruinous loss (measured by the investment percentage of total assets) and impact on liquidity (measured by payback period). Accola argues that these are different aspects of risk which should be separately evaluated, and he proposes the AHP as a means of combining the three individual 'scores', hence generating a single measure of project risk. Finally, Accola suggests that intangible benefits be taken into account, again using the AHP in order to produce a project 'score' for those benefits specific to the project.

The outputs of the framework are three groups of values for each decision alternative: return on investment, risk and non-quantifiable uncertainties. The model does not recommend a specific alternative but the outputs are represented in a three-dimensional graph to determine the efficient plane which includes all alternatives that dominate the other alternatives.

Accola's proposals are logical, grounded in existing theory and recognise the importance of not duplicating the same item(s) several times in the analysis. Nevertheless, there are three main criticisms on Accola's proposal. First, he does not justify the three aspects of risk which should be considered, he simply asserts them. Second, he treats intangible benefits and project 'uncertainty' as synonymous. This leads him to the paradoxical conclusion that increasing project 'uncertainty' is good. It would be simpler to treat the 'uncertainty' dimension of his model as representing intangible benefits and so avoid any confusion. Third, in developing his model, Accola chooses product reliability, product performance, flexibility and employee benefits as intangible benefits. As in the risk dimension, he does not justify why these particular benefits and not others are included in the model.

Assessment of the Analytic Hierarchy Process The AHP is more sophisticated than simple scoring models because it recognises the importance of superordinate and subordinate criteria. Like simpler models, the AHP requires subjective weightings of criteria at each level of the hierarchy so that they may be combined into a single metric at the next higher level. However, although the weightings are undeniably subjective, the consistency checking process ensures that weightings are internally consistent. Thus, the AHP provides an attractive framework for modelling complex, multi-attribute, decision problems. The hierarchical analysis is intuitively sound and internal consistency checks on the relative weightings of criteria increase confidence in the model's reliability. The AHP may also be more preferable than scoring models because the decision maker compares only each pair of factors/ alternatives to determine which of them is more important, without any obligation to set the absolute weight of each factor/ alternative (Azzone et al., 1992).

However, the AHP is not without its pitfalls. There is no theoretical framework for structuring the problem into a hierarchical representation (Boucher and MacStravic, 1991). The same problem may be represented as different hierarchies by different decision makers, and consequently they can arrive at different conclusions for the same problem. The pairwise comparison matrices might be time consuming, managers may lose interest and most fundamentally, as Boucher and MacStravic (1991) pointed out, the kinds of questions asked in making pairwise comparisons may be meaningless. What does it mean to ask how strongly one prefers lead time reduction to quality improvement? For the decision maker to be able to interpret the question, he or she needs to know how much lead time reduction is being compared to how much quality improvement. The assumption is that the decision maker must be thinking of some average quantities when answering the question, otherwise it would not be reasonable to make a judgement. Also, there is a flaw in the method of aggregating relative weights into composite relative weights because the weights at one level of the hierarchy cannot be determined independently of weights on another level of the hierarchy.

Discussion and Conclusions

In attempting to overcome the difficulties of applying traditional investment decisions based on ROI and cash flow analysis, three approaches have been suggested in the normative literature of AMT investment decision making. Some researchers advocate that AMT investment decisions should be based on the results of traditional cash flow analysis. Intangible benefits should be considered only if the NPV is negative. This approach was suggested by Kaplan (1986) in response to his question 'Must CIM be justified on faith alone?' Basing his argument on the fundamental soundness of DCF analysis, Kaplan seems to suggest that the answer is 'no'; like other investments CIM can be rationally justified. However, if a negative NPV is calculated, Kaplan suggests that intangible benefits might be considered in order to 'make up the deficit', but here it seems that Kaplan is advocating a decision based on 'faith': faith in the intangible benefits. Another suggestion in this tradition is that of Park and Son (1988). They tried to quantify intangible benefits: quality and flexibility, and include them in a multi-period NPV linear programming model. Leaving aside the difficulty of putting financial values on these benefits, the model is limited by the assumptions of linear programming.models.

Some researchers have suggested a completely opposite approach: recommending that investment in AMT be regarded as a strategy that should be implemented regardless of the results of cash flow analysis. Meredith and Suresh (1986) suggested that the evaluation of fully integrated systems should be based on strategic grounds. They suggested four criteria as the basis for the strategic analysis: technical importance, business objectives, competitive advantage and research and development. Medearis et al. (1990) and Elango and Meinhart (1994) argued that FMS could allow firms to improve their competitive position, basing their analysis on the strategic forces identified by Porter (1980). We believe that proposals in this tradition underestimate the importance of financial analysis for AMT investments. Normally, a range of projects compete for investment funds and it is necessary to identify the optimum areas to deploy capital. The problem can be acute when considering AMT systems with options ranging from stand-alone systems to fully integrated systems (Primrose and Leonard, 1986). The 'strategic' approach leads to neglect of important financial information and depends greatly on subjective judgements that may not be very helpful in ranking several alternatives. Also there is a limit to which such a strategic approach can be adopted without compromising the financial position of the firm.

While some researchers have advocated that decisions should be based on quantifiable factors and others have suggested that AMT investment be driven by strategic considerations, a third group of researchers have attempted to reconcile these approaches by building models which can accept both economic (financial) and strategic (intangible) variables. The simpler models are based on a simple combination of multiple scores so as to generate an overall project 'score'. Three methods of scoring have been suggested: unweighted 0-1 scoring (Meredith and Suresh, 1986), unweighted factor scoring (Nelson, 1986) and weighted factor scoring (Parsaei and Wilhelm, 1989). These scoring models are attractive because they allow both strategic and financial variables to be combined. However, the models suggested are sensitive both to the factors included in the model and to the relative weight subjectively assigned to them. Also, the lack of a consistency measure means that the AHP has advantages over simple scoring methods.

A number of researchers have adopted the AHP in modelling AMT investment decision making, aiming to integrate both qualitative and quantitative variables in a single model. This integration may be achieved in one of four ways: screening the alternatives on the basis of their financial performance and then choosing amongst the screened alternatives according to an analysis of their strategic performance, strategic analysis is based on the AHP (for example: Naik and Chakravarty, 1992); screening the alternatives on the basis of their strategic performance (again derived from the AHP) and then choosing amongst the screened alternatives according to their financial performance (for example: O'Brien and Smith, 1993); modifying the results of DCF methods by weighting according to the qualitative variables, and then choosing the best alternative according to the modified DCF (for example: Srinivasan and Millen, 1986); or considering the financial performance as a factor or dimension in a multi-dimensional framework (for example: Putrus, 1990 and Accola, 1994).

Reviewing previous models of AMT investment decision making indicates that the problem of measuring and incorporating project risk into the decision model has not been given the same attention as the problem of quantifying intangible benefits. Only a few studies have taken this problem into consideration (for example, Kulatilaka, 1984; Miltenburg and Krinsky, 1987; Putrus, 1990; Accola, 1994). Two approaches have been considered: the moment-oriented approach and the dimension-oriented approach. The moment-oriented approach requires data such that the multiple outcomes of the various alternatives and their related probabilities can be estimated. According to this approach, the riskiness of an investment project may be measured by its standard deviation or variance. The dimension-oriented approach is based on analysing a project's riskiness into dimensions and is usually used when the probability distribution is difficult to estimate (Accola, 1994).

Adjusting the discount rate to reflect the riskiness of a project (a moment-oriented approach) has been criticised because there are many different determinants of risk and it is difficult to capture all the aspects of project risk through a single modification of the discount rate (Ronen and Sorter, 1972), even if this adjustment is based on a sophisticated technique such as the capital asset pricing model. Kaplan and Atkinson (1989, p. 479-480) criticise the capital asset pricing model's 'beta' in this adjustment because it reflects for only one type of risk: the systematic risk of shareholders (that risk which is not diversifiable across firms). Moreover, they object to accounting for this type of risk because it probably arises more from the nature of the firm's product markets than from uncertainties in its production processes, and therefore, it would be the same for both existing and proposed investments in process technology.

Criticisms of the methods of adjusting the discount rate in order to account for AMT project risk, and the difficulty of estimating probability distributions for these, often novel, investments can lead to a preference for the dimension-oriented approach. It is argued that by adopting this approach managers can evaluate projected cash flows separately from the risks associated with them (Accola, 1994).

Reviewing the normative literature of AMT investment decision making leads to the conclusion that an appropriate framework should provide a systematic consideration of both qualitative variables and quantitative variables. Slagmulder et al. (1995) summarised the prevailing view: 'more and more authors are convinced that good investment appraisal requires that strategic and financial considerations be reconciled and integrated.' (See, also, for example: Lefley, 1996, p. 347 and Currie, 1994, p. viii.) The model developed in later chapters of this book is based on an analysis of both the qualitative variables *and* the financial evaluation of projected cash flows. Additionally, following Ronen and Sorter (1972) and Accola (1994), the framework considers project risk separately from cash flow projections. Accordingly, the three key dimensions of financial return, non-financial benefits, and risk form the basis of the model.

The authors contend that the model has a strong theoretical base, and is developed in a manner which addresses two basic criticisms that can be levelled against almost all previous models. First, previous models are based on techniques which require decision makers to provide *precise* values for all factors included in the decision model. Many of these factors are qualitative, any scores assigned to them are subjective, and so, the required precision cannot be achieved. Second, there is insufficient justification for the inclusion of some factors and not others in the decision model. Both of these criticisms may lead to difficulties in applying previous models in practice.

Fuzzy set theory is suggested as a means of overcoming the first criticism, the required precision in estimating factors in the decision model, as fuzzy set theory was developed primarily to deal with such problems. Decision makers no longer need to provide precise estimates for the decision factors, rather they are allowed to model the vagueness and ambiguity (fuzziness or inexactness) in the decision model[7]. Thus, the question now becomes '*how can fuzzy set theory be utilised to develop an investment model that enables decision makers to provide estimates of decisional factors as **imprecise** values or even as **verbal*** statements?' This question is addressed in chapter 7.

The second criticism regarding the lack of sufficient justification of the factors included in the decision model is tackled by conducting an empirical study. From this study the key factors affecting the decision to invest in AMT can be deduced and hence included in the decision model. Decision makers are likely to have more confidence using a model that is driven from practice. The design and methodology of the empirical study is discussed in the next chapter.

Notes

1 Many classification schemes have been suggested in this field. For example, Meredith and Suresh (1986) categorised the evaluation techniques of investment in advanced manufacturing technology into three separate approaches: economic, analytic and strategic. Soni et al. (1990) classified the economic justification methods for AMT into three classes: single-objective deterministic methods, multi-objective deterministic methods, and probabilistic/stochastic methods. Swamidass and Waller (1990) reviewed the literature on the acquisition of new manufacturing technologies. They broadly divided the literature into planning and financial justification categories. The financial justification category was divided into six approaches: DCF, scoring, costs and benefits, risk analysis, computer enhanced, and strategic value of flexibility. Finally, Kolli et al. (1992) proposed another classification scheme in which the economic justification methods were classified into two major groups: single criterion and multiple criteria methods. Each major group was divided into deterministic and non-deterministic methods.

2 All of these methods are well documented in almost all text books available on the topic of management accounting or engineering economics (see, for example, Lumby, 1995; Garrison, 1991; Hilton, 1991; Riggs and West, 1986; and Stevens, 1979).

3 To determine the required amount of flexibility, first, Elango and Meinhart (1994) defined strategic flexibility as a set of flexibilities available in the operational units of an organisation which facilitate and support the manufacturing strategy. Then, they categorised the flexibility of manufacturing into three types: design flexibility, volume flexibility, and product mix flexibility. Each type was defined and measured as follows. Design flexibility is the ease with which new parts can be added or substituted for existing parts and the degree to which the part mix currently being produced can be changed inexpensively and rapidly. It can be measured by: cost or time required to shift from one part mix to another, number of new parts that can be introduced every year with a given system, and/or benefits of product flexibility in terms of incremental revenue accrued to the firm due to a certain value of design facility. Volume flexibility is the ability of a manufacturing system to be operated profitably at different overall output levels and it has two aspects: speed of response and range of variation. Volume flexibility can be measured by the stability of manufacturing costs over widely varying levels of total production volume. Product mix flexibility

is the range of goods that the manufacturing system can produce without adding major capital equipment. It may be measured by the range of diversity the system is capable of handling.

4 Nelson used the subscript i to refer to the shop with respect to the equipment evaluation and capacity elasticity. But, in other variables, i, refers to the project. Nelson recognises this inconsistency in his work (1986, 347).

5 To obtain the normalised pairwise comparison matrix, first sum the values in each column in the pairwise comparison matrix, then divide each element in the column by its column total.

6 The Leontief model is used to determine total sector multipliers. Each total sector gives the expected number of transfers for an item within the sector (manufacturing centre). Linear programming is used to determine the expected maximum number of both finished items and scrap.

7 The usefulness of fuzzy set theory to reduce the need for precise estimation of the decision factors is discussed in chapter 6.

3 Empirical Research Design

In the previous chapter it was stated that researchers can investigate AMT investment decision making by either a normative, 'theoretical', approach or a descriptive, 'empirical', approach. One might expect that theoretical models for AMT decision making based on a three-dimensional framework of 'return-risk-intangible benefits', would overcome the apparent shortcomings of traditional investment models based on ROI or cash flow analysis. However, most previous models can be criticised for two reasons. First, they employ techniques which require precise values for all input variables. The difficulty of achieving this precision suggests fuzzy set theory as a technique which can accept estimates of the decision factors, whether expressed in approximate numerical values such as 'approximately £1000', and/or in linguistic values such as 'high'. Second, previous models were usually built on hypothetical situations in which an arbitrary set of factors were assumed to affect the final decision. Almost all of these models lack sufficient justification for the inclusion of some factors and the exclusion of others.

The second approach for investigating AMT investment decision making is the descriptive approach. Here practitioners are either surveyed or interviewed in order to draw conclusions regarding the actual practice of investment decision making. Although many empirical studies have investigated the practice of capital investment decisions in general, very few studies have been directed at investigating AMT investments in particular.

This chapter is divided into four sections:

1. A review of previous empirical studies.
2. The objectives of the empirical study and research questions.
3. Research methodology.
4. Conclusions.

A Review of Previous Empirical Studies

During the last two decades many empirical research studies (for example: Klammer et al., 1991 in the USA, Pike and Wolfe, 1988 in the UK, Van Cauwenbergh et al., 1996 in Belgium, and Kalyebara, 1996 in Australia) have addressed the issue of company capital budgeting theory and the practices used by practitioners. These research studies have examined topics such as capital budgeting techniques, assessment of project risks, estimation of cash flows and discount rates, and the relationship between various techniques used in evaluating and selecting investment projects on the one hand, and firm performance, project type, manager's performance, firm characteristics and decision-making effectiveness on the other (see for example: Klammer, 1973; Schall et al., 1978; Pike and Roberts, 1980; Kim et al., 1986; Pike, 1988a; 1988b; Hendricks, 1988; Meredith, 1989; Pike et al., 1989; Currie, 1989; Boaden and Dale, 1990; Klammer et al., 1991; Meredith and McCutcheon, 1991; Ho and Pike, 1991; 1992; Butler et al. 1991, 1993; Lefley, 1994a, 1994b; Carr et al., 1994; Chen and Small, 1994; Nixon, 1995; Collier and Gregory, 1995; Chen, 1995; Shao and Shao, 1996; Shank, 1996; and Carr and Tomkins, 1996).

Such studies can be categorised into two groups: the first, larger, group investigated investment projects in general and the second, smaller, group investigated AMT investment projects in particular. The first group will be briefly reviewed in the next section, then the second group, concentrating on AMT investments, will be considered in more detail.

Empirical Studies Regarding Investment Decisions in General

The practice of investment appraisal techniques such as payback, ROI, and DCF methods, and risk analysis techniques such as sensitivity analysis, adjustment of the discount rate, and adjustment of the payback period, have been major subjects in almost all previous surveys. Such techniques have often been classified into sophisticated techniques, such as DCF methods in investment appraisal and sensitivity analysis/management science techniques in risk analysis, and unsophisticated (or naive) techniques, such as the payback and ROI as investment appraisal techniques and adjusting required payback period or required ROI as risk analysis techniques, (see, for example, Klammer et al., 1991; Pike, 1988a; Ho and Pike, 1991; and Chen, 1995). Accordingly, previous surveys have attempted, first, to describe the state of the art of sophisticated and unsophisticated practices, second, to examine if there are relationships between the use of sophisticated techniques and firm performance, effectiveness of decision making, type of investment projects, or firm characteristics, and third, to investigate how hurdle rates are determined in practice. A brief review of some of the main surveys is presented in table 3.1.

Generally, previous empirical studies of investment decision practices indicate some common findings. For example, surveys have reported the increasingly widespread use of sophisticated capital budgeting techniques such as NPV and IRR to evaluate investment projects, and with a preference of IRR over NPV. However, the payback period method is still the most widely used evaluation technique with virtually all companies using it either as a primary or secondary technique. Perhaps practitioners understand the limitations of sophisticated methods so continue to employ the payback method alongside more sophisticated methods. Surveys also show that sensitivity analysis has become a popular method for analysing the riskiness of investment projects.

Examination of the association between capital budgeting techniques and company characteristics has indicated that sophisticated capital budgeting techniques are most likely to be used by larger companies. In addition, the use of sophisticated capital budgeting techniques may contribute to capital budgeting effectiveness. However, surveys have failed to show a consistent significant association between firm performance and the sophistication of the capital budgeting techniques used.

Empirical Studies Regarding AMT Investment Decisions

As stated in chapter one, AMT investments have special characteristics such as the huge amount of investment required, the expected long life of the sophisticated systems, their potential benefits (which tend to be qualitative in nature), and the lack of experience of using high technology systems, which makes them risky projects. While many theoretical approaches have been suggested[1], little attention has been given to investigating the practice of decision making for such investments.

Two studies have surveyed AMT investment decisions in UK practice. The first study was conducted by Pike (1989) who surveyed AMT investment decisions in a sample of 47 large UK firms. The sample included 19 process industry firms, 17 general manufacturing firms, and 11 non-manufacturing firms. The study reported the importance of 16 factors in arriving at investment decisions. The factors were divided into four

Table 3.1 Previous studies regarding investment decisions in general

Author(s)	Sample	Main issues	Main conclusions
Klammer (1973)	148 US industrial firms	Investigation of whether firm-level performance is related to the sophistication of capital investment procedures and standards used.	Results of regression analysis did not show a consistent significant association between performance and capital budgeting techniques
Schall et al. (1978)	189 US firms	Capital budgeting techniques employed, computation of the discount rate and cash flows, and the method of estimating and adjusting for risk. Examination of the association between firm characteristics with the sophistication of capital budgeting techniques.	A trend toward more sophisticated capital budgeting techniques. Required rate of return was based on an after-tax weighted average cost of capital. Slight evidence that sophistication in capital budgeting methods is positively related to the size of the firm's capital budget and negatively related to the firm's beta value.
Pike (1988a)	100 large UK firms	Examination of the trend towards greater sophistication in investment selection techniques and control processes between 1975 and 1986, and their resulting impact on the effectiveness of evaluating and controlling capital projects.	Very significant increases in the sophistication of capital budgeting over the period. Regression analysis indicated that more sophisticated techniques were significantly associated with higher levels of capital investment effectiveness.

Klammer et al. (1991)	100 large US industrial firms	Comparison of techniques over three interval periods: 1980, 1984 and 1988 for various types of projects: replacement, expansion, foreign operations, etc.	Limited changes in the use of project evaluation and management science techniques but an increased use of formal risk analysis.
Chen (1995)	115 large US manufacturing firms	Investigation of the use of project evaluation techniques across three types of projects: equipment replacement, expansion of existing products and expansion into new products. Examination of the association of firm characteristics (from three perspectives: agency theory, contingency theory and financial aspects) and capital budgeting techniques used.	DCF models more important than payback method, and payback more important than accounting rate of return in all three types of investments. Non-financial techniques as important as DCF models in evaluating both equipment replacement and expansion into new products. DCF models receive high importance in firms that give high importance the objective of maximising stockholders' wealth (agency perspective). The impact of contingency theory variables on capital budgeting techniques has had mixed results. Profitability (ROA) did not have an impact on the use of capital budgeting techniques, while debt ratio and firm size were significantly related to the use of capital budgeting techniques. Firms with high use of DCF methods were borrowing significantly more than firms with low use of DCF, and were significantly larger in their capital expenditures.

categories: strategic, quantifiable performance, qualitative performance, and risk factors. It was found that process industry firms assigned a highest average weight to strategic factors, followed by non-manufacturing firms with the lowest weight being assigned by general manufacturing firms. This suggested that general manufacturing firms placed much less emphasis on strategic factors than did the other two groups of firms. Also, it was noted that the four most important factors for process industries were: 'degree of fit with business strategy', 'growth rate of market related to the project', 'sensitivity to change in key assumptions', and 'effect on product quality'. For general manufacturing the four most important factors were: 'sensitivity to change in key assumptions', 'degree of fit with business strategy', 'payback period', and 'effect on product quality'. For non-manufacturing firms the factors were: 'degree of fit with business strategy', 'sensitivity to change in key assumptions', 'effect of achievement of profit and sales targets' and 'investment track record of project sponsor'. However, the evidence did not indicate statistically significant differences between such factors, either within the same group or among the three groups of firms.

The second study, conducted by Lefley (1994a), surveyed AMT investment practice in a sample of 134 large UK manufacturing firms. Three main issues were investigated: financial appraisal methods used, assessment of risk, and methods of management appraisal. It was found that the most popular method of investment appraisal was the payback method which was used by 93.8% of the respondents while only 69.4% of respondents used either IRR or NPV. This could indicate a decline in the use of more sophisticated methods. However, such results are contrary to the findings of almost all previous surveys. One could hypothesise that AMT investment is evaluated mainly by naive methods, but this hypothesis should not be accepted without more supporting evidence.

Again the payback method was the most popular method of assessing risk in AMT investments (71.5% of respondents). The second most popular method was sensitivity analysis (56.3% of respondents). Again, this is contrary to earlier findings which had indicated an increasing use of sensitivity analysis (see for example: Ho and Pike, 1991). Regarding methods of management appraisal, this survey indicated that 73.6% of firms were using the 'net profit before tax basis' and 59.7% were using 'cash flow' as a means of assessing managerial performance. This means that most companies used short-term measures in management appraisal. Lefley concluded that:

> '... on the whole a managers' performance is assessed using some form of short-term performance criteria, although there is some indication that companies are taking a long-term strategical approach to this assessment ...' (p. 2772)

This may provide a reason for claims that AMT investments are lacking in UK industry. It should be noted that the results of this survey, like Pike's (1989) survey results, have not been subjected to statistical testing in order to investigate any relationship between investment appraisal techniques and managers' performance measures. Neither did Lefley's survey investigate the non-financial issues in AMT investment decision making.

The dominance of questionnaire surveys in empirical research in investment decision making has recently been criticised by Nixon (1995). He argued that the case study approach offers a better opportunity to examine in-depth, the many variables that motivate or constrain investment in new technology. His field study was based on six companies (three US companies and three UK companies) in order to evaluate the relative influence of management accounting practice, including capital investment appraisal techniques, on decisions to invest in new technology. Nixon found that three issues influenced AMT investment decisions in the six companies in varying degrees. They were: the consequences of deferring investment, risks and the degree of strategic consistency. Examination of these three issues highlighted the technical complexity of the decisions and the wealth of information from many internal and external sources that was sought in order to evaluate likely cash flows, risks and strategic implications. In addition to the technical complexity of AMT projects, Nixon also found that the investment decisions had a 'political' context: they had implications that were long term and affected many individuals and interest groups both inside and outside of the company. In each case the process of investigation, discussion and negotiation took over a year involving a whole series of sub-decisions before a final commitment was made. Nixon concluded that once the decision process is better understood the 'information which is already impounded in these decisions in a relatively unstructured, informal, qualitative and subjective way can be moved towards a more structured, formal, quantitative and objective domain' (p. 285).

Unlike Nixon who did not distinguish between US and UK companies, Currie (1994) conducted a cross-national comparison for managing AMT in manufacturing industry. This study was based on semi-structured interviews with managers in eight Japanese companies, five US companies, six West

German companies, and five UK companies. Generally, the results of this study are consistent with that of Lefley's (1994a) survey. The UK and the US companies demonstrated a greater propensity to evaluate AMT using short-term performance targets (3 years or less) than was the case for their West German and Japanese counterparts. In the UK and US, companies' career and reward systems were based on the attainment of short-term performance targets while Japanese companies placed a high premium upon length of service and loyalty. Currie concluded her study with a call for developing new methods of evaluating AMT in which a wider array of financial and non-financial benefits could be considered. Such methods could improve management's understanding of some of the advantages of AMT and supplement traditional investment models based on cash flow analysis.

Slagmulder and Bruggeman (1992) investigated the investment analysis and justification processes of investment in FMS in six Belgium manufacturing companies, basing their research on semi-structured interviews with plant managers and FMS project managers. They hypothesised that 'different methods of investment justification (such as payback and DCF methods) can lead to high financial performance of investment projects in FMT (flexible manufacturing technology), but each method is not equally effective under all conditions'. They concluded that quality of strategic analysis is a critical success factor in the decision regarding investment in FMT. In other words, an investment justification based on sound strategic arguments appears to be a necessary condition for success. Although all companies in the sample performed a detailed DCF or payback calculation for their investments, the outcome of the financial analysis was not considered to be the key decision criterion in all cases. Some companies went ahead with a project even if the calculated return was lower than the hurdle rate.

The Objectives of the Empirical Study and Research Questions

Although the practice of AMT investment decision making has been heavily recommended for further investigation (for example: Ho and Pike, 1991, p. 236; Klammer et al, 1991, p. 130), limited research has been undertaken to date. It will be recalled that one of the criticisms of the AMT decision models described in chapter 2 was the ad hoc selection of variables included in the models. Given the relative lack of studies of AMT investment decision making in particular and of studies which consider non-financial factors in general, a survey was undertaken in order to provide a grounded basis for the variables to be included in the decision model that will be developed in chapter 7.

The main aims of the empirical study were to explore current practices for evaluating investment decisions regarding the adoption of AMT, and to identify the major factors affecting these decisions in practice. In particular, the following research questions were addressed:

1. What are the current levels of various types of AMT in large UK manufacturing companies?
2. How important are the financial return measures used in evaluating capital investment projects?
3. Which specific financial return measures are preferred?
4. How important are the non-financial investment criteria used in evaluating capital investment projects?
5. How important are the risk analysis techniques used in evaluating capital investment projects?
6. Are there differences between evaluating AMT investment and other types of investment? If so:
7. In which areas do these differences exist?
8. How can the benefits of AMT investment be considered?
9. What are the opinions of financial directors of large UK manufacturing companies regarding difficulties experienced in AMT decision making?
10. What sorts of risk are encountered in AMT investments? If any:
11. How can these risks be measured and incorporated in the investment decision?

Research Methodology[2]

Unlike previous empirical studies, the study reported here was based on two methods of research: a cross-sectional survey and a field study. We believe that this 'triangulation' provided a better understanding of AMT investment practices by minimising the limitations of the separate research methodologies. The cross-sectional survey aimed to collect quantitative data using a structured questionnaire. Five point Likert scales facilitated the quantification of responses so that statistical analysis could be undertaken and differences between groups of companies could be observed and generalised. This methodology is thus based on the positivistic approach and objectivity can be achieved, to some extent, through using well-known concepts when constructing the questions and providing definitions for concepts which may not be so familiar to respondents. The field study aimed to collect qualitative data in order to describe how an AMT investment decision was made in specific companies, and to provide some illustrations of the survey results. This methodology is based mainly on the phenomenological approach. The researchers thus analysed the qualitative data using an indexing system (discussed later) and by producing extensive quotations from the interview transcripts. In addition, the case reports were approved by the interviewees and, in some cases, modified according to the interviewees' comments. This approach to analysing and interpreting the field study results means that a positivistic approach dominated the whole research process even when qualitative data were collected and analysed.

Triangulation of research methodologies can enhance the validity and reliability of the research results (Ghauri et al., 1995, p. 93), and can minimise the limitations of each research methodology (Sarantakos, 1993, p. 155). Moreover, multiple methods of data collection can maximise the amount of data collected (Easterby-Smith et al., 1991, p. 133). However, some authors (for example: Sarantakos, 1993 and Easterby-Smith et al., 1991) warn that the use of triangulation per se is not necessarily more valuable than a single-method approach for certain research questions. Taking this warning into consideration, triangulation in this research was adopted for two main reasons. First, a large sample of companies was needed in order to reach generalisable conclusions regarding the key factors that affect AMT decisions, and to identify differences between AMT decision making and other types of investment project. For this, a cross-sectional survey based on a mail questionnaire was considered to be the most appropriate research methodology. Second, further empirical enquiries, such as an explanation of how AMT investment decisions can be made by reference to specific examples, could not be achieved effectively using a cross-sectional survey. Hence, a field study based on the case study approach was undertaken. Major issues related to these two methodologies are discussed next.

The Cross-sectional Survey

The current survey differed from previous surveys in several respects. First, most previous surveys have not investigated qualitative factors affecting investment decisions. When qualitative factors have been considered, they have been treated as one group (Chen, 1995), or only a few factors have been investigated (Pike, 1989). This study reports the importance of most of the qualitative factors reported in the literature on the evaluation of investment decisions.

Second, most previous surveys asked respondents to categorise capital budgeting and risk analysis techniques as of primary or secondary importance. In this survey respondents were asked to rate the importance of each technique or criterion on a five-point equal distance rating scale. This type of question is designed to give a better understanding of the relative importance of each technique in arriving at investment decisions. Since previous surveys indicated that most companies use more than one technique to evaluate capital investment projects, it is logical to gather information about the importance of each technique rather than to categorise or rank them. Moreover, ranking is only practical for a limited number of techniques, while in this survey, eleven non-financial criteria were included. However, to provide information comparable with previous surveys, respondents were asked to identify which financial appraisal technique was regarded as most important if more than one technique was used.

Third, as noted by Chen (1995), most previous surveys reported only simple descriptive statistics and no statistical comparisons were made. In this study, Kruskal-Wallis one-way ANOVA was used to compare the importance of various techniques and criteria used in the evaluation of investment projects across each group of companies studied. Kruskal-Wallis one-way ANOVA is a non-parametric statistical test for repeated measures and three or more mutually independent samples (see, for example, Gibbons, 1993 and Siegel and Castellan, 1988)[3]. The matched parametric test is not applicable because the five-point scale used in rating the importance of investment techniques and criteria is an ordinal measurement scale and parametric assumptions are not satisfied.

Finally, while previous studies have compared results by project type (replacement, expansion, etc.), little attention has been given to distinguishing between AMT and other types of investment project. In this study AMT investment projects were compared with non-AMT investment projects. In addition, the degree of integration of AMT was considered by classifying AMT projects into two groups: less integrated projects (such as CNC, CAM, CAD and robots), and fully integrated projects (such as CIM and FMS).

Data Collection Method A postal questionnaire was used to collect primary data. This method was chosen to facilitate access to a large number of respondents and thus provide sufficient data for statistical analysis. Such a large number would be difficult to achieve through face-to-face interviews because of the constraints of time and cost. In addition, the pressures on financial directors would probably have restricted the number willing to be interviewed. However, a limited number of interviews were conducted; first, to pilot the questionnaire before sending it out, and second, to follow up the questionnaire in order to check the reliability of the questionnaire results.

Three interviews were undertaken between September 1995 and January 1996. The first interview explored issues to be included in the questionnaire. In the other interviews, interviewees were asked to answer the questions with comments concerning their clarity and relevance. Two interviews were tape recorded and each took approximately one and a half hours. The third interview was with an academic member of staff who had long experience in industry. After improvements, four academic staff were asked to comment on the questionnaire design. After a very few final refinements, the questionnaire was sent to the sample companies in the last week of January 1996.

Nine interviews with questionnaire respondents were conducted between April and June 1996 with key personnel responsible for investment decisions in their companies. Five were face-to-face interviews and four were conducted by telephone. These interviews served three purposes: to check the reliability of questionnaire responses; to investigate some issues raised by the questionnaire results further; and to investigate other issues which could not be investigated in the questionnaire. The design of these interviews is discussed in the field study section.

Sample Frame and Size To achieve the stated aims of the survey, companies included in the sample had to have invested in AMT. It was believed that the most appropriate sample frame would be composed of the largest UK manufacturing companies. After reviewing a number of available databases, it was found that the 'FAME' (Financial Analysis Made Easy) was the most appropriate from which to choose the sample frame. FAME is a CD-ROM database which includes extensive information about 160,000 major public and private British companies[4].

In selecting companies for inclusion, two criteria had to be satisfied: they were large (in terms of turnover, number of employees, and fixed assets), and they operated in the manufacturing sector. These two criteria

Table 3.2 Stages in selecting the companies to be included in the sample

Stage	Criterion used	No. of companies
1	SIC UK primary industry code (3 and 4)	30,522
2	Minimum turnover of 30 million pounds	6,920
3	Minimum number of 100 employees	20,985
4	Minimum fixed assets of 30 million pounds	4,577
5	Boolean expression (stages 1, 2, 3 and 4)	578
6	Independent companies	559

were achieved by carrying out the following six stages, the results of which appear in table 3.2:

1. The companies were chosen according to their primary SIC UK industry codes of first digit '3' (metal goods, engineering and vehicles industries) and first digit '4' (other manufacturing industries).
2. 'Large' companies were identified by stipulating a minimum turnover of £30 million for the last year (1994)[5].
3. A further test for 'largeness' was inserted by stipulating a minimum of 100 employees.
4. Because, ideally, the targeted companies would have invested in AMT companies with at least £30 million of fixed assets were chosen.
5. The previous four stages were undertaken independently, that is, the number of companies quoted satisfies only the criterion applied at that stage. When criteria were combined, 578 companies were identified to satisfy them all.
6. So as to assure the independence of the companies included in the sample, all the companies' addresses and lists of directors were inspected. When two or more companies were found at the same address and have the same directors, only one company was selected. This reduced the sample to 559 companies, which was judged to be a reasonable size.

One of the major disadvantages of mail surveys is the generally poor response rate (for example, Newell, 1993; and Frankfort-Nachmias et al., 1992). As an attempt to achieve a better response rate, a letter was sent to all companies in the sample prior to sending the main questionnaire. A letter addressed to the financial directors of the 559 companies was sent in the last week of November 1995. The objective of the letter was to obtain pre-agreement for completing the questionnaire. It contained two parts: the first part served to introduce the study and its general aims, in addition to defining the terminology of AMT, and emphasising the confidentiality of the information supplied by each respondent. In the second part, the respondent was asked to answer four questions[6]. The first question regarded the willingness of the respondent to participate in the survey. The second and third questions were to obtain information regarding the company and the respondent. The last question was to rank the importance of the main three issues in the survey (financial payoff, risk analysis, and the non-financial criteria) for both AMT and other types of investment projects[7].

Three letters were returned by the post office stamped 'addressee has gone away' leaving 556 potential responses. By the end of January 1996 a total of 139 responses were received giving a response rate of 25%. Surprisingly, of these responses, 90 respondents were not willing to participate in the survey: 19 because of company policy, 27 because either the company had not invested in AMT or lack of time, while 44 did not state any reason.

The number of companies that agreed to participate in the survey was thus 49. This number was judged to be a rather small sample from which to elicit conclusions regarding the practice of investment decisions. Further inspection of the companies which had not given a reason for non-participation in the survey (44 companies), revealed that most of them had actually answered all the questions in the letter. This was a motivation to explore why they did not agree to participate in the survey. Four of them were chosen at random and called by telephone[8]. All four considered the letter to be a survey in itself and were not willing to spare more time for the research[9].

In the light of this, it was decided that all companies which had not replied to the letter (417 companies) would be sent the questionnaire in addition to the companies which agreed to participate in the survey (49 companies). Accordingly, 466 questionnaires were sent to the financial directors at their companies' addresses. Further, it was thought that addressing financial directors by their names, rather than by their job title, would motivate them to respond to the questionnaire. Some of the financial directors' names were available in the letter (49 companies) and in 'FAME' (64 companies). The remaining companies (353 companies) were telephoned to obtain the name of their current financial directors. In total the financial directors' names for 222 companies were obtained while in the other companies either the switch-board person said that s/he was not permitted to give it, or the telephone number for the company was wrong. Accordingly, 131 questionnaires were addressed to the attention of the 'financial director'.

The Questionnaire Format A covering letter was included as part of the questionnaire[10]. The objective of this letter was to explain the purpose of the study, to explain to the respondent the importance of answering the questions, and to assure the complete confidentiality of the information provided by the respondent. Moreover, by using window envelopes and sticking the address label on the covering letter, the respondent to the questionnaire could be identified. This was essential in order to make follow-up telephone calls, and it also facilitated obtaining secondary data from the FAME database.

The questionnaire requested two types of information, factual and attitudinal. Most of the questions were closed, typically yes/no or five-point equal distance scale with two extreme end points, for example, not important/extremely important, or strongly disagree/strongly agree. The questions were grouped together in the questionnaire according to subject and type of information requested so that it was possible to format the questions in a matrix (Frankfort-Nachmias et al., 1992).

The questionnaire (reproduced in Appendix 3) was composed of 14 questions. Questions 1 and 3 aimed to identify the involvement of respondents in the evaluation of investment projects and whether the firm had invested in AMT. These questions also served as filters. Questions 2 and 11 aimed to elicit the personal opinion of the respondent regarding the extent to which difficulties were experienced in the estimation of the basic elements of the cash flow calculations, and to identifying the extent to which the intangible aspects of AMT could provide obstacles to its implementation. Questions 4 and 5 were specific questions designed to determine the types of AMT projects that the company had invested in and the expected life of each. Question 4 also aimed to ascertain the level of AMT in each company and the pattern of investment in each type of AMT. Question 6 was the central question in the survey. It aimed to identify the importance of four groups of techniques and criteria used in evaluating investment decisions: financial appraisal techniques, non-financial criteria, risk analysis techniques, and three possible dimensions of risk. In questions 7 and 8 the respondents were asked to identify the required payback period most frequently used, and the range of minimum rates of return or discount rates required by the company. Question 9 identified the financial appraisal technique which the company relied on most in cases where more than one technique was used. Question 10 ascertained how the firm considered the benefits of AMT projects: in financial terms, non-financial terms, or not at all. Questions 12 and 13 asked the respondent if s/he was willing to participate in interviews and whether s/he would like a copy of the survey results. Finally, question 14 was used to collected demographic data regarding the respondent and his/her company.

Responses to the Questionnaire 466 questionnaires were sent to the financial directors of the sample of large UK manufacturing companies in the last week of January 1996. The questionnaires were addressed either to the financial director by name (335 companies) or by official job title (131 companies). In the last week of February, follow-up telephone calls were made to all non-respondents. By the end of March, 138 questionnaires were received giving a response rate of 29.6%. Of those received, 36 were returned unanswered because the company policy was not allowed to respond to surveys (20 companies), the questionnaire was not relevant to the company (8 companies), lack of time (5 companies), or the company was in the middle of a merger (3 companies). The net response rate was thus 23.7% (102 completed questionnaires / 430 potential respondents). Further, three completed questionnaires were judged as not valid to the analysis because the respondents had not participated in the evaluation process of any investment project. Hence, 99 usable completed questionnaires were used in the analysis giving a net usable response rate of 23% (99/430). This response rate is comparable with other similar surveys such as those of Lefley (1994a) and of Chen (1995) which had response rates of 28.8% and 20% respectively. Such response rates are acceptable because of the pressure under which the financial directors work in present economic circumstances, and because of the large number of questionnaires that they are asked to complete. For example, one financial director stated that:

> '... I have completed numerous questionnaires and been interested in the results when they are published. However, we are now receiving so many requests from students, universities, business schools and other organisations who are conducting increasingly extensive surveys, that it is becoming absolutely impossible to devote enough time to properly complete the questionnaires to make the responses of value ... I have, therefore, decided to decline all invitations to assist in such projects ...'

Evaluation of Non-response Bias The main threat to the conclusions of a postal survey is non-response bias. In order to assess the possibility of the existence of such bias, the 99 responding companies and the sample frame of 466 companies were compared. This comparison was based on the two criteria used to choose the original sample: industry distribution (according to the first digit of SIC UK primary industry), and size of the company measured according to turnover, fixed assets, and number of employees for the year ended 1994.

Two types of statistical test were used to compare the two samples of responding companies and sample frame. First, a non-parametric Chi-square test was used to test the homogeneity of industry distribution. Second, parametric two-sample t-tests were conducted on company size measured by turnover, fixed assets, and number of employees for the year ended 1994. In comparing the actual sample with the sample frame, there is a null hypothesis that 'there is no significant difference between the two samples'. Accepting the null hypothesis means that, most probably, the two samples are drawn from the same population (see for example, Cramer, 1994; Kvanli, 1992; and Sirkin, 1995) and it can be concluded that existence of non-respondents is not likely to be a threat to the conclusions based on the actual sample.

Table 3.3 provides the results of the statistical tests for each of the four measures. The null hypothesis for each is that there is no difference between the two samples.

The results of the Chi-square test on the industrial distribution shown in table 3.3 suggests that the first null hypothesis is rejected because of the small p-value (.005). The crosstabulation for industry distribution of both samples suggests that responding companies might be biased towards SIC UK industry code '3'. The t-tests for the turnover, fixed assets, and number of employees resulted in the rejection of the second null hypothesis and acceptance of the last two null hypotheses. These results indicate that the responding companies are significantly larger than the sample frame in terms of turnover (p-value = .007), but are not significantly different in terms of fixed assets (p-value = .943) or the number of employees (p-value = .981). It can be concluded that the responding companies are more representative of companies with first digit of SIC primary code '3' (metal goods, engineering, and vehicle industries) and larger companies in terms of turnover for the year ending 1994.

To investigate the effect of non-response bias, the answers to the main questions in the questionnaire from respondents who further replied without

Table 3.3 A summary of statistical tests

	Responding companies[a]	Sample frame[b]	Statistical comparisons
1. Industrial Distribution (SIC)			
First digit 3[c]	58 companies (58.6%)	201 companies (43.1%)	
First digit 4[d]	41 companies (41.4%)	265 companies (56.9%)	
Pearson Chi-square value			7.85350
P-value			.00507
2. Turnover (1994)			
Mean	602,075	386,980	
Standard deviation	1,140,125	578,962	
t-value			2.690
P-value			.007
3. Fixed assets			
Mean	182,441	185,783	
Standard deviation	321,998	425,830	
t-value			-0.070
P-value			.943
4. Number of employees			
Mean	5,894	5,933	
Standard deviation	11,128	14,618	
t-value			-0.020
P-value			.981

[a] 99 responding companies.
[b] 466 companies.
[c] Metal goods, engineering and vehicles industries.
[d] Other manufacturing industries.

the follow-up telephone calls (72 respondents) were compared with the answers from respondents who replied only after the follow-up telephone calls (27 respondents). There was no significant difference between the two groups of answers. For example, mean differences in the importance of 'payback', 'quality and reliability of outputs' and 'sensitivity analysis' were 0.12 (p-value = .883), 0.10 (p-value = .834) and 0.20 (p-value = .318). Thus, it could be concluded that more respondents would not change the results of the study.

From the above, it can be concluded that any non-response bias has a very small effect, if any, and is not likely to be a threat to the conclusions of the study. However, any bias would be likely to mean that the findings are more generalisable to larger rather than smaller companies, and to companies with SIC UK primary code '3' rather than '4'.

Data Analysis Most questions were either categorical questions or scoring questions. In the categorical questions, respondents were asked to choose between a set of items. Answers to these questions were presented as frequency tables. In second type, scoring questions, respondents were asked to assign a score on a five point equal distance scale. Answers to these questions were presented as both frequency tables and as tables of descriptive statistics.

Responding companies were categorised into three groups: those not employing AMT, those using less integrated forms of AMT, and those using fully integrated AMT. This categorisation aimed to identify differences between the evaluation of AMT and other types of investment, and to investigate any relationship between the sophistication of investments themselves and the sophistication of the technique(s) used to evaluate them. In order to facilitate statistical comparisons between these groups, simple descriptive statistics were computed for the importance of various investment techniques and criteria across each group of companies. The Kruskal-Wallis one-way ANOVA test was then used to determine whether there were statistically significant differences between the AMT investment behaviour of the three groups of companies.

The Field Study

The term field study is often used to describe studies of social practices in the field of activity in which they take place (Ryan et al., 1992, p. 113). Frankfort-Nachmias (1992, p. 284) state that 'the goal of a field research is to develop a theory that is "grounded," that is, closely and directly relevant to the particular setting under study.' Given that the purpose of the research was to develop an empirically based model for the evaluation of AMT investment decisions incorporating fuzzy set theory, a field study was seen as an appropriate methodology to ensure that the model was 'grounded' in practice. Possible approaches to field studies include field experiments, action research and case studies (Sangster and Baldwin-Morgan, 1996). Both field experiments and action research require an intervention, and in most cases, some manipulation by the researcher. A descriptive case study approach (Yin, 1989) was adopted in the current field study in order to develop the required 'descriptive' model of AMT investment decision making.

Sarantakos (1993, p. 263) gave several justifications of the case study approach: it provides information on specific cases, after or before large-scale studies (such as surveys) are undertaken, it illustrates larger studies by offering insights into specific aspects of these studies, and it offers supplementary information. These three reasons provide the motivation for undertaking a field study. For example, the field study enabled the further investigation of issues raised by the survey results, such as how practitioners manage risk in AMT investment projects, reasons for choosing specific financial return measures, and how financial and non-financial factors could affect the final decision. Recently, a similar approach, following up a survey by case studies in order to generate insights, was adopted by leading researchers such as Scapens et al. (1996).

The usual criticism of the case study approach is its lack of generalisability. Rayan et al. (1992) distinguished between two types of generalisability that arose from empirical research. These are: 'statistical generalisation' and 'theoretical generalisation'. Statistical generalisation is concerned with statements about statistical occurrence in a particular population. Theoretical generalisation attempts to generalise theories so that they explain the observations that have been made. As a result of the small sample involved, statistical generalisation is usually impossible for case studies. They can, however, be used as a basis for theoretical generalisation.

Two basic decisions should be taken before undertaking a case study. First, what is the unit of analysis? Second, how should access to the case site be gained? Rayan et al. (1992, p.113) state that 'a case study, however, usually implies a single unit of analysis.' In the field study here, the unit of analysis in each case was 'an AMT investment decision'. Second, in order to gain access to companies that had experience of AMT investment decision making, respondents to the questionnaire were asked whether they were willing to participate in interviews. Fourteen respondents answered 'yes' to this question. Five of these respondents were excluded because they did not have experience in AMT investment decision making, and the remainder were contacted by telephone. Five of the respondents agreed to face-to-face interviews[11] and the other four agreed to be interviewed by telephone. Table 3.4 provides some demographic information of the interviewees. The

Table 3.4 Some demographic information of the interviewees

Company	Job title	Background	Highest degree or qualification	Experience in the company	Nature of the interview
A	Finance director	Accounting & Finance	ACMA (1990)	5 years	Face-to-face
B	Finance manager	Accounting & Finance	MA (1987) ACA (1990)	2 years	Face-to-face
C	Finance director	Accounting & Finance	FCA (1962)	4 years	Face-to-face
D	Finance controller production	Engineering & Finance	MBA (1991)	4.5 years	Face-to-face
E	Group technical director	Science	PhD (1973)	10 months	Face-to-face
F	Business analyst	Accounting & Finance	FCA (1978)	5 years	Telephone
G	Finance director	N/A	MA (1968)	5 years	Telephone
H	Financial planning manager	Accounting & Finance & Engineering	CA (1992)	1 year	Telephone
I	Finance director	Accounting & Finance	BA (1977)	4 years	Telephone

A: Baby Food Ltd., **B**: Engineering Plc., **C**: Newspaper Plc., **D**: Tobacco Ltd., **E**: Packaging Plc., **F**: Heavy Engineering Plc., **G**: Maritime Ltd., **H**: Spirit Ltd, **I**: Shipbuilding Ltd.

names of the companies have been disguised in order to maintain confidentiality.

The interviews were semi-structured using a list of open ended questions prepared in order to maintain consistency between the interviews. However, the list of questions was not introduced at the start of each interview; instead, each interviewee was asked to describe how an investment decision for a recent major AMT project had been made. By the end of the interview, most of the prepared questions were answered by the interviewee. However, the listed questions were checked with the interviewee in case any had not been answered.

The interview schedule included four sets of questions. The first set was designed to obtain general information regarding the main motivations for investing in AMT, the process of evaluating investment projects in the company and the evaluation team. The other three sets of questions were designed to elicit the main factors considered in the evaluation of the investment project: financial, non-financial and risk factors[12].

All the interviews were tape recorded and then transcribed verbatim. The non-verbal illustrations in the face-to-face interviews were noted during the interview. The transcriptions and notes together with publicly available data for the companies were studied and analysed using QSR NUD.IST[13]. In order to use this computer package, each interview was transcribed into a word-processed document. When entered into NUD.IST this document could be divided into text units. Then an index system was created by categorising each text unit under one or more headings, known as nodes, by the program. In this way, each interview was re-structured into a number of nodes. Each node included all the text units relating to a specific issue discussed in the interview. This method of analysing qualitative date allowed the researchers to *maximise* objectivity both with analysing the qualitative data generated from the interviews and with representing each interview as a case study. Each case synopsis set out in chapter 5 describes and summarises how a specific investment decision was made and the main factors that affected the decision.

Each written case study was sent to the interviewee with a covering letter in which he was asked to comment on the content of the case, its presentation, and its relation to actual company practice. Only three interviewees requested a few amendments. The cases resulting from the face-to-face interviews and two telephone interviews are presented in chapter 5. Two other telephone interviews proved difficult to transcribe because of the quality of the recordings. However, these two cases confirmed the results reported and they are reflected in the discussion and conclusions of the field study.

Conclusions

In this chapter previous empirical studies of investment decision have been reviewed. Although numerous models have been suggested in the normative AMT decision making literature, few empirical studies have been undertaken. The survey conducted by Pike (1989) revealed the importance of non-financial factors in AMT investment decisions but only a few non-financial factors were considered, the sample was small (47 companies) and included 11 non-manufacturing companies. The survey conducted by Lefley (1994a) indicated the popularity of the payback method as both a financial appraisal technique and a method of risk assessment. Lefley argued that the short-term criteria used when assessing managers' performance can be a reason for using short-term criteria (such as short payback periods) in the evaluation of investment projects. This short-termism could be a reason for relatively slow take-up of AMT investments in UK industry. Neither survey was supported by statistical tests and there were no comparisons between AMT investments and other types of investment. Although the literature has indicated the importance of non-financial issues in AMT investment decision making, Lefley did not investigate this, and Pike included only a few non-financial factors in his survey.

Field researchers such as Nixon (1995), Currie (1994) and Slagmulder and Bruggeman (1992) have investigated AMT investment practice using the case study approach. In general their studies highlighted the importance of strategic factors in evaluating such investments. While cross-sectional surveys can provide a means of collecting data from a large sample of companies and this can facilitate comparisons between companies, field studies can provide rich data which describe and allow insights into different practices. Thus, triangulation using the two types of research methodology was achieved in the current study. Through using triangulation, the advantages of both types can be maximised and their limitations can be minimised. In the following two chapters, the results of the cross-sectional survey and the field study are presented.

Notes

1 See chapter 2.

2 In social science, it is argued that there are two approaches from which research methodology can be derived: a positivistic approach and a phenomenological approach (see for example: Easterby-Smith et al., 1991, ch. 2; Creswell, 1994, ch. 1; and Hussey and Hussey, 1997, ch. 3). The most significant distinguishing feature between the two research approaches is that adoption of either approach leads to specific research methodologies. The objectivity of the positivistic approach requires a research methodology that is concerned with collecting and analysing quantitative data in order to arrive at generalisable inferences, which are often based on statistical analysis. Contrariwise, the phenomenological approach requires a research methodology that is concerned with collecting and analysing qualitative data in order to describe and understand a phenomenon in its context. However, the choice of either methodology should be based on the research questions which are to be addressed.

It should be noted that the positivistic and phenomenological approaches are two extremes, and it is unlikely that researchers in the social sciences operate in either of their pure forms. Although the basic beliefs of the two approaches may be quite incompatible, the differences between the actual research methodologies adopted by the researchers are not clear cut. Easterby-Smith et al. (1991) indicated that there is an increasing move amongst management researchers to develop methods and approaches which provide a middle ground, with some bridging between the two extreme approaches.

3 In the computation of the Kruskal-Wallis test, for a specific variable, all scores of all groups are ranked in a single series. The lowest score is replaced by rank 1, the next lowest score is replaced by rank 2 and the largest score is replaced by rank N, where N is the total number of independent respondents in all the groups. The sum of ranks in each group is then calculated and divided by the total number of respondents within each group to obtain the average rank. The Kruskal-Wallis test assesses the differences between the average ranks to determine whether they are significant or not. SPSS for windows 6.1 was used to analyse the survey data.

4 The FAME database is a computerised service provided by CD-ROM publishing Co. Ltd., 1 Great Scotland Yard, London SW1A 2HN, and Jordan & Sons Ltd., 21 St Thomas Street, Bristol BS1 6JS. This service was obtained through the computer network available at University of the West of England, Bristol.

5 It should be noted that the figure of £30 million was chosen arbitrarily after many attempts to narrow the sample to around 500 companies. This figure was based on an expected response rate of usable completed questionnaires of 20%.

6 A copy of the letter is reproduced in Appendix 1.

7 A pre-paid self-addressed envelope was enclosed.

8 It was not reasonable to call all of the 44 companies given their desire not to participate in the full survey.

9 If the process were to be repeated, certainly, the decision would be not to send a preliminary letter.

10 A copy of the cover letter is appended in Appendix 2.

11 One of these face-to-face interviews was conducted at the university in response to the interviewee's request.

12 Appendix 4 contains the interview schedule.

13 QSR NUD.IST stands for Non-numerical Unstructured Data Indexing Searching and Theorizing. It is a computer package designed to aid users in handling non-numerical data in qualitative analysis (see, QSR NUD.IST user guide, 1995).

4 The Cross-sectional Survey of UK Investment Decision Making

The cross-sectional survey was undertaken in order to explore current practice of capital investment decision making with particular emphasis on the adoption of AMT. The objective was to identify the key factors affecting such decisions, hence the following research questions were investigated:

1. What are the current levels of different types of AMT in large UK manufacturing companies?
2. How important are the financial return measures used in evaluating capital investment projects?
3. Which specific financial return measures are preferred?
4. How important are the non-financial investment criteria used in evaluating capital investment projects?
5. How important are the risk analysis techniques used in evaluating capital investment projects?
6. Are there differences between the evaluation of AMT investments and other types of investment? If so,
7. In which areas do these differences exist?
8. How can the benefits of AMT investment be considered?
9. What is the opinion of financial directors of large UK manufacturing companies regarding difficulties experienced in AMT decision making?

From 466 questionnaires mailed to the financial directors of large UK manufacturing companies 99 usable completed questionnaires were received and used in the analysis.

Summary and Analysis of the Survey Results

The survey results are presented under the following major headings:

1. Demographic information.
2. Level of AMT investment in the sample companies.
3. Responses by AMT companies.
4. Responses by both AMT and non-AMT companies

Demographic Information

Some demographic information relating to respondents was collected. Most of the respondents (90%) were either finance directors or financial controllers, and the remainder were treasurers, technology specialists, or investment analysts. Most respondents had an accounting and finance background (83%); and the remainder had backgrounds in engineering (7%), operations (5%), or mathematics, science, and social science (5%). Accounting and finance professionals are more likely to be involved in developing and evaluating capital investment projects than people in other functional areas (Farragher, 1986 and Chen, 1995). Most respondents had been in their positions for 2 years or more (72%), and 63% of respondents have been with their current company for 5 years or more. All respondents have been involved in the evaluation of investment decisions and 78% had experience of AMT investment projects. Respondents were generally well informed about investment decisions within their companies.

Level of AMT Investment

The term AMT was used in the questionnaire to refer to any form of computerised manufacturing process. Seven types of automated manufacturing systems, frequently referred to in the literature were listed: CNC, robotics, AMH, FMS, CAD, CAM and CIM. Meredith and Suresh (1986) categorised these systems into three groups according to the level of integration of the system. The first level comprises stand-alone systems such as CNC and robotics; the second level comprises linked systems such as CAD, CAM, and AMH; and the third level comprises integrated systems such as FMS and CIM.

Table 4.1 Investment in AMT by type of projects

	No	%
Computer Aided Design (CAD)	65	86
Computer Numerical Control (CNC)	48	63
Automated Material Handling (AMH)	43	57
Computer Aided Manufacturing (CAM)	42	55
Robotics	33	43
Flexible Manufacturing Systems(FMS)	32	42
Computer Integrated Manufacturing (CIM)	27	36

Respondents were asked to indicate whether their companies had invested in AMT or not. Seventy-six companies (77%) had invested in at least one form of AMT. These respondents were asked to specify the type(s) of AMT they had invested in. Table 4.1 shows the number and percentage of companies investing in each form of AMT. The list is not mutually exclusive, and several companies had invested in more than one form of AMT. The majority of firms had invested in CAD, with CNC coming second. The lowest percentages were for FMS and CIM, which might support the view of Wilner et al. (1992) that high level of technology investments are implemented incrementally rather than on a fully integrated basis.

Table 4.2 and figure 4.1 show the periods in which the AMT projects were implemented. It may be concluded that fully integrated systems have been installed relatively recently. For example, about 59% of CIM

Table 4.2 Types of AMT projects according to the installation period

	Pre 1985		1985-90		1991-96	
	No	%	No	%	No	%
Computer Aided Design	11	17	30	46	24	37
Computer Numerical Control	15	31	15	31	18	38
Automated Material Handling	6	14	15	35	22	51
Computer Aided Manufacturing	5	12	16	38	21	50
Robotics	9	27	14	43	10	30
Flexible Manufacturing Systems	4	12	13	41	15	47
Computer Integrated Manufacturing	3	11	8	30	16	59

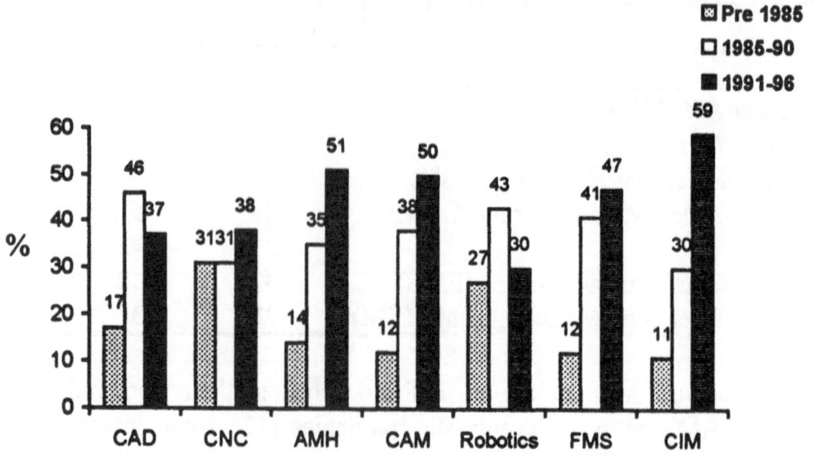

Figure 4.1 Types of AMT projects according to their installation period

investments were undertaken during the period 1991-96 while more than 60% of the stand-alone systems such as CNC and robotics were implemented before 1991. This is not surprising if companies invest in automation in incremental stages rather than in a revolutionary fashion.

In order to investigate whether there were differences between the methods used in investment evaluation and the different levels of AMT, companies were classified according to the level of AMT adopted. There were no companies in the first level of AMT (stand-alone level), so companies were classified into three groups: non-AMT companies, less integrated companies which have invested in one or more of the first five types of AMT listed in table 4.1 but not in FMS or CIM, and fully integrated companies which have invested in FMS and/or CIM. Table 4.3 shows the number and percentage in each group.

Lefley (1994a) reported a relative lack of investment in AMT by UK manufacturing companies with only 43% of the companies in his study either having implemented or being in the process of implementing integrated AMT systems[1]. The corresponding percentage in the current survey is almost the same (43.4% of companies invested in integrated AMT systems), still a low percentage bearing in mind that the sample is drawn from the largest UK companies.

Table 4.3 Classification of companies according to their level of AMT

	No	%
Non-AMT companies	23	23.2
Less integrated AMT companies	33	33.3
Fully integrated AMT companies	43	43.4

Responses by AMT Companies

The questionnaire contained four groups of questions relating only to AMT investments. Thus, only respondents who had experience in evaluating AMT investment projects (76 respondents) could answer these questions. The questions concentrated on the life of AMT investments, the benefits of AMT, difficulties experienced in the evaluation process, and the opinion of respondents regarding certain propositions.

Life of AMT Investments Information on the expected life of AMT projects was obtained. As shown in table 4.4 and figure 4.2, the majority of companies expected a life of 4-6 years for CAD and CAM (56% and 51% respectively). For the other types of AMT project, the expected life was longer. For example about 87% of companies expected a life of 7 years or more for CNC and about 73% expected the same life for robotics. The difference in expected life between CAD/CAM and other types of AMT

Table 4.4 The expected life for AMT projects

	3 years or less		4-6 years		7-10 years		more than 10 years	
	No	%	No	%	No	%	No	%
Computer Aided Design	13	20	36	56	11	17	4	7
Computer Numerical Control	2	4	4	9	34	72	7	15
Automated Material Handling	2	5	11	27	26	63	2	5
Computer Aided Manufacturing	4	9	22	51	14	33	3	7
Robotics	1	3	8	24	22	67	2	6
Flexible Manufacturing Systems	2	7	8	27	16	53	4	13
Computer Integrated Manufacturing	1	4	10	38	12	46	3	12

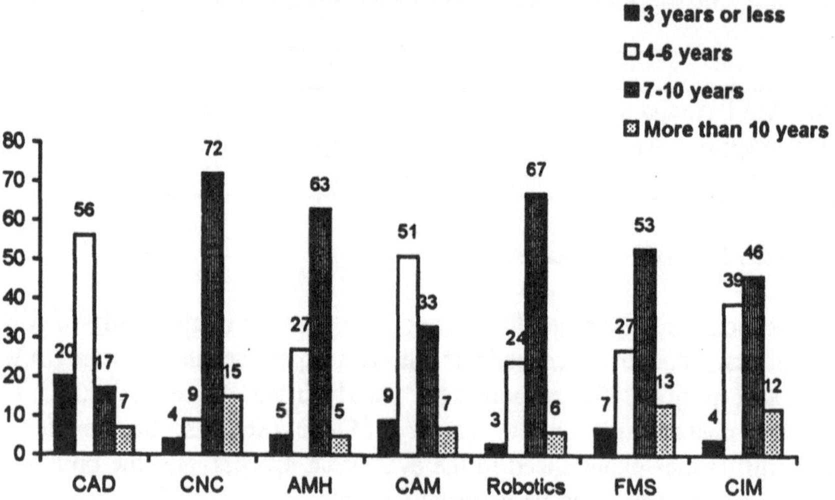

Figure 4.2 The expected life of AMT projects

might be attributed to the faster obsolescence of software compared to hardware technology.

Benefits of AMT Investments In the literature, it is usually reported that firms adopting AMT projects can gain numerous benefits. Twenty-one benefits were listed and respondents were asked to identify whether each benefit was considered financially, non-financially, or not at all in the evaluation process. The results are shown in table 4.5 and figure 4.3.

These results indicate that eight benefits were always considered, either financially or non-financially, in evaluating AMT investments: reduced material costs (100%), reduced scrap and rework costs (100%), improved product quality (100%), reduced labour costs (98%), reduced inventories level (98%), faster response to market needs (97%), improved competitive position (97%), and reduced lead times (96%). Conversely, a significant minority did not consider the following four benefits in the evaluation process, either financially or non-financially: obtaining experience of new technology (45%), effects on employee morale (40%), floor space reduction (34%), reduced after-sale costs (32%).

Table 4.5 Benefits of AMT investments (% of respondents)

	Considered financially	Considered non-financially	Not considered at all
Reduced labour costs	97	1	2
Reduced material costs	96	4	-
Reduced inventory levels	92	6	2
Reduced scrap and rework costs	89	11	-
Increased sales volume	79	8	13
Savings from less frequent set-ups	69	23	8
Increased manufacturing capacity	52	38	10
Improved product quality	14	86	-
Faster response to market needs	11	86	3
Consistency with corporate strategy	10	78	12
Improved competitive position	20	77	3
Greater manufacturing flexibility	17	75	8
Reduced lead times	22	74	4
Improved company image	-	71	29
Easier production scheduling	12	65	23
Retention of market share	22	58	20
Increased market share	28	54	18
Reduced after sale costs such as warranty	41	27	32
Floor space reduction	27	39	34
Effects on employee morale	1	59	40
Obtaining experience of new technology	6	49	45

Most firms were willing to quantify seven benefits in financial terms. These benefits were: reduced labour costs (97%), reduced material costs (96%), reduced inventory levels (92%), reduced scrap and rework costs (89%), increased sales volume (79%), savings from less frequent set-ups (69%), and increased manufacturing capacity (52%).

The benefits which were considered in the evaluation process on a non-financial basis by the majority of respondents were: improved product quality (86%), faster response to market needs (86%), consistency with corporate strategy (78%), improved competitive position (77%), greater manufacturing flexibility (75%), reduced lead times (74%), improved company image (71%), and easier production scheduling (65%).

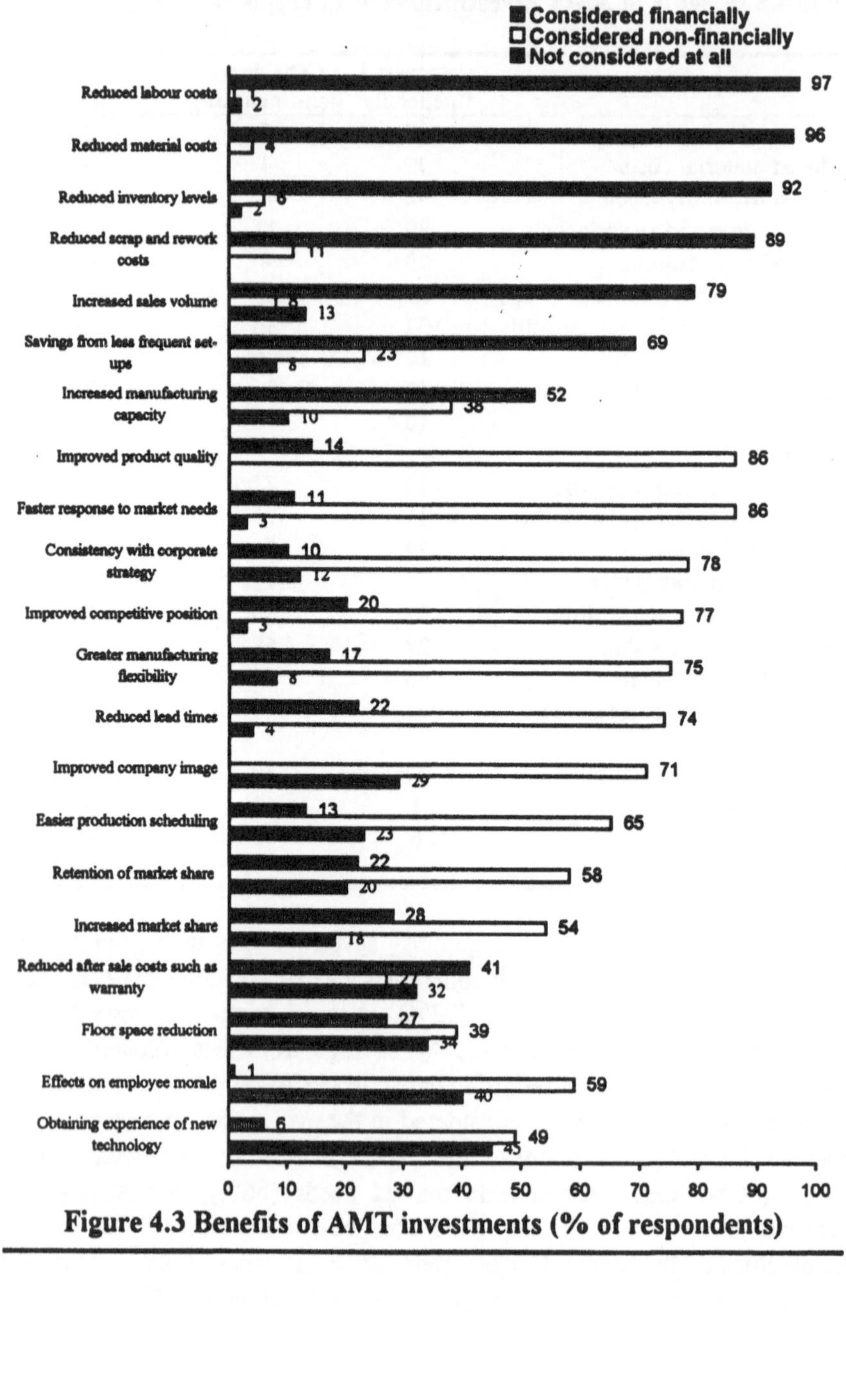

Figure 4.3 Benefits of AMT investments (% of respondents)

Table 4.6 Difficulties in evaluating AMT investment projects

	% of respondents					Mean	Std. dev.
	Score 1	Score 2	Score 3	Score 4	Score 5		
Quantifying intangible factors	-	2	9	46	43	4.29	0.71
Assessing risk	2	8	32	43	15	3.63	0.89
Estimating energy savings	9	18	33	28	12	3.16	1.14
Estimating project life	-	27	40	30	3	3.09	0.83
Predicting cash flows	-	35	46	12	7	2.91	0.88
Estimating material savings	5	55	24	13	3	2.55	0.89
Estimating labour savings	7	56	28	6	3	2.41	0.83

Score 1: Very easy, Score 2: Easy, Score 3: Neutral, Score 4: Difficult, Score 5: Very Difficult

Difficulties in Evaluating AMT Investment Projects On a five-point equal distance scale ranging from 'very easy' to 'very difficult', respondents were asked to rate seven key elements in the evaluation of AMT investment projects. The percentages of respondents at each point on the scale are shown in table (4.6). It was found that an overwhelming number of respondents (89%) rated 'quantifying intangible factors' as either 'difficult' (score 4) or 'very difficult' (score 5). Also, 'assessing risk' was considered by 58% of respondents as a 'difficult' or 'very difficult' task in the evaluation process. On the other hand, 'estimating labour savings', and 'estimating material savings' were judged to be relatively easy tasks by the majority of respondents, 63% and 60% respectively. Finally, a significant minority thought that 'predicting cash flows', 'estimating project life', and 'estimating energy savings' were neither 'easy' nor 'difficult' tasks in the evaluation process (46%, 40%, 33% of respondents respectively).

Opinions of Respondents for Three Propositions Respondents were asked to indicate to what extent they agreed or disagreed with three propositions relating to financial evaluation of AMT investments using a five-point equal distance scale ranging from 'strongly disagree' to 'strongly agree'. The results are given in table (4.7). Only 15% of respondents agreed or strongly agreed with the first proposition: 'it is difficult to get AMT investment proposals approved because of stringent financial criteria'. The second proposition: 'some potential benefits of AMT are not taken into account

Table 4.7 Opinions of respondents regarding three propositions

	% of respondents					Mean	Std.
	Score 1	Score 2	Score 3	Score 4	Score 5		dev.
a. It is difficult to get AMT investment proposals approved because of stringent financial criteria.	15	34	37	12	3	2.54	0.97
b. Some of the potential benefits of AMT are *not* taken into account because they are difficult to express in financial terms.	3	22	24	46	5	3.29	0.96
c. It is possible to implement an AMT investment proposal based *only* on potential non-financial benefits.	41	32	12	12	4	2.07	1.17

Score 1: Strongly disagree, Score 2: Disagree, Score 3: Neutral, Score 4: Agree, Score 5: Strongly agree.

because they are difficult to express in financial terms' received considerable support from 51% of respondents who agreed or strongly agreed with this statement. Finally, a large number of respondents (73%) either disagreed or strongly disagreed with the last proposition 'it is possible to implement an AMT investment proposal based only on potential non-financial benefits'.

Responses by Both AMT and Non-AMT Companies

The questionnaire contained some questions directed at all respondents regardless of whether their companies invested in AMT or not. These questions were about the importance of various aspects of investment decision making. These were classified into three groups: financial appraisal techniques, non-financial criteria, and risk analysis techniques. Respondents were also asked to indicate their company's required payback and discount rate.

Financial Appraisal Techniques Financial appraisal techniques such as return on investment (ROI), payback period (PB), discounted payback period (DPB), net present value (NPV), and internal rate of return (IRR) have usually been surveyed by asking respondents to rank these techniques as primary or secondary methods (see for example: Lefley, 1994a; Klammer et al., 1991; Pike and Wolfe, 1988). In this survey, respondents were asked to rate the importance of each technique used for evaluating investment projects on a five-point equal distance scale. Table 4.8 shows the percentage of respondents at each point on the scale.

Table 4.8 The importance of the financial appraisal techniques (% of respondents)

	Score 1	Score 2	Score 3	Score 4	Score 5
Return on Investment	6	6	13	30	45
Payback	3	2	20	38	37
Discounted Payback	14	16	16	35	19
Net Present Value	7	13	21	25	34
Internal Rate of Return	6	7	18	30	39

Score 1: Not important at all, Score 2: Not important, Score 3: Neutral, Score 4: Important, Score 5: Extremely important.

Table 4.8 shows that all five techniques are used in most companies and all five techniques are judged as important (score 4) or extremely important (score 5) by most companies. Most respondents consider ROI and payback methods as either important or extremely important (75%) while 69%, 59% and 54% of respondents consider IRR, NPV and discounted payback respectively as either important or extremely important.

Respondents who use more than one technique were asked to specify the technique that was most important in evaluating investment projects. Table 4.9 and figure 4.4 show the number and percentage of respondents who consider a particular technique to be most important.

Table 4.9 The most important financial appraisal technique

	No	%
Return on Investment	20	21
Payback	28	30
Discounted Payback	6	6
Net Present Value	14	15
Internal Rate of Return	26	28
Total	94	100

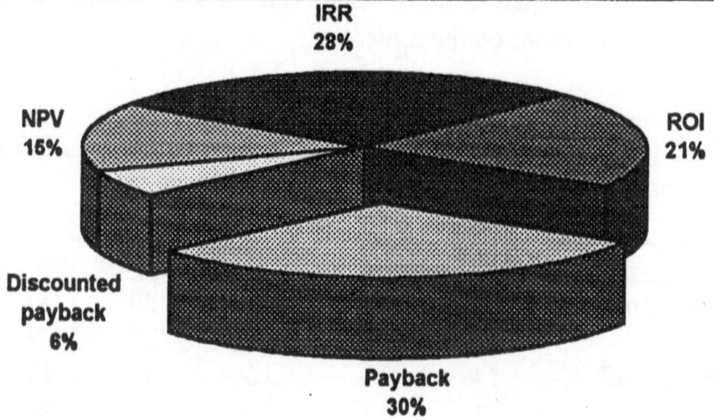

Figure 4.4 The most important financial appraisal technique (%)

Table 4.9 reveals that a total of 94 companies (95%) use more than one financial appraisal technique for investment projects. Of these companies, 30% preferred payback over the other techniques while 28% preferred IRR. The least preferred method was discounted payback. Most companies (57%) consider one of the so-called naive methods (ROI and payback methods) to be most important while the remainder (43%) consider one of the sophisticated methods (NPV and IRR) as most important with preference for IRR rather than NPV (IRR is preferred by almost twice as many companies as NPV).

The results shown in both table 4.8 and table 4.9 give a clear indication of the continuing importance of the payback method in the evaluation of investment projects. However, more sophisticated analysis is needed in order to reach conclusions about investment decision making specifically in relation to AMT.

Analysis of the Financial Appraisal Techniques According to Level of AMT

The sample was classified into three groups according to the level of AMT in each company: 'non-AMT companies', 'less integrated companies', and 'fully integrated companies'. Descriptive statistics are summarised in table 4.10 for each financial appraisal technique used in each group of companies.

Cursory investigation of table 4.10 reveals that all five techniques play an important role in evaluating investment projects, and that there is little variation across the three groups of companies. For every technique some firms responded with 1 (not important) and others with 5 (extremely important) except for the payback method in the second group (minimum score 3) and for IRR in the third group (minimum score 2). The median for all five techniques is 4 except for ROI in the third group (median 5). There is some variation in mean importance of the five financial appraisal techniques. ROI, payback, and IRR are most important with different ranking across the three groups of companies. In the first group, ROI is most important and payback comes second followed closely by IRR. In the second group, the payback method comes first and ROI and IRR are second and third respectively. Finally, in the third group, ROI again is first followed by payback and IRR. In all three groups NPV comes fourth and discounted payback is the least important technique. It is difficult to reach any conclusions regarding the importance of each financial appraisal technique

Table 4.10 Descriptive statistics for the financial appraisal techniques

	Mean	Std. dev.	Median	Min.	Max.
Non-AMT Companies					
Return on Investment	4.09	1.11	4	1	5
Payback	3.73	1.20	4	1	5
Discounted Payback	3.10	1.34	4	1	5
Net Present Value	3.50	1.36	4	1	5
Internal Rate of Return	3.70	1.42	4	1	5
Less integrated Companies					
Return on Investment	3.83	1.31	4	1	5
Payback	4.23	0.77	4	3	5
Discounted Payback	3.42	1.28	4	1	5
Net Present Value	3.55	1.27	4	1	5
Internal Rate of Return	3.78	1.31	4	1	5
Fully integrated Companies					
Return on Investment	4.17	1.07	5	1	5
Payback	4.06	0.96	4	1	5
Discounted Payback	3.31	1.37	4	1	5
Net Present Value	3.84	1.21	4	1	5
Internal Rate of Return	4.06	0.94	4	2	5

across the three groups of companies without conducting a more sophisticated statistical analysis.

To investigate differences in the importance of the financial appraisal techniques between the three groups of companies, Kruskal-Wallis one-way ANOVA was used. Table 4.11 shows the results of these tests.

Examination of table 4.11 reveals that there is no significant difference between the three groups for any of the five financial appraisal techniques under investigation, the probability levels being very insignificant. This leads to the conclusion that the importance of each of the financial appraisal techniques is not affected by the nature of the projects being evaluated.

Table 4.11 Results of Kruskal-Wallis test regarding financial appraisal techniques

	Mean Rank			Kruskal-		
	Non AMT	Less integrated	Fully integrated	Wallis Statistic	D.F.	P-value
Return on Investment	44.11	39.83	46.16	1.1827	2	.5536
Payback	40.39	50.65	46.75	2.1210	2	.3463
Discounted Payback	37.17	42.38	41.21	0.6614	2	.7184
Net Present Value	40.67	41.17	46.85	1.2582	2	.5331
Internal Rate of Return	39.67	40.20	43.54	0.5006	2	.7786

Non-financial Criteria

Recently, considerable attention has been given to the evaluation of investment projects in general and AMT investment projects in particular, relating to non-financial criteria (see, for example, Elango and Meinhart, 1994; Medearis et. al., 1990; and Mensah and Miranti, 1989). These criteria are sometimes referred to as non-traditional evaluation techniques as opposed to the (traditional) financial evaluation techniques (Chen, 1995). Typically, they refer to key non-financial dimensions such as quality of products, consistency with corporate strategy, and current and potential future competition. As indicated by Chen (1995, p. 148): 'these factors are considered to be critical in the new manufacturing environment and portray the competitive advantages of a project which cannot be easily assessed by the traditional financial-based techniques'.

In order to investigate the role played by such 'non-financial criteria' when evaluating investment projects, respondents were asked to evaluate 11 criteria using a five point equal distance scale. Table 4.12 gives the percentage of respondents at each point on the scale.

It can be seen from table 4.12 that all 11 criteria are used in most companies but with considerable divergence in the level of importance attached to each. Eighty percent or more of respondents considered the criteria 'quality and reliability of outputs', 'requirements of customers', and 'consistency with corporate strategy' to be important or extremely important. The criteria 'obtaining greater manufacturing flexibility', and 'reduced lead-times' were rated as important or extremely important by 78% and 75% of respondents respectively. However, the last three criteria in the

Table 4.12 The importance of the non-financial criteria (% of respondents)

	Score 1	Score 2	Score 3	Score 4	Score 5
Quality and reliability of outputs	1	1	11	43	44
Requirements of customers	1	3	10	40	46
Consistency with corporate strategy	2	4	14	45	35
Obtaining greater manufacturing flexibility	1	2	19	46	32
Reduced lead-times	6	3	16	41	34
Keeping up with competition	6	6	19	43	26
Reduced inventory levels	4	9	20	37	30
The ability to expand in the future	3	4	35	42	16
Improved company image	9	19	39	30	3
Employee morale	4	29	43	22	2
Experience with new technology	14	18	45	21	2

Score 1: Not important at all, Score 2: Not important, Score 3: Neutral, Score 4: Important, Score 5: Extremely important.

table were rated as important or extremely important by relatively few respondents 33%, 24% and 23% respectively.

Analysis of Non-financial Criteria According to Level of AMT Descriptive statistics for each non-financial criterion were calculated and these are shown in table 4.13. Examination of this table provides an indication of the variation in importance of non-financial criteria in evaluating investment projects in the three groups of companies. For example, the criterion 'quality and reliability of outputs' is most important in the third group of companies but it is fourth in the first group. Likewise, 'obtaining greater manufacturing flexibility' is the third criterion in both the second and third groups though it is the seventh in the first group. However, there is some degree of pattern in the responses. For example, the criterion 'requirements of customers' is one of the two most important criteria in all three groups of companies. Also, the criteria 'improved company image', 'employee morale', and 'experience with new technology' are the least important criteria in all three groups.

To analyse the importance of each non-financial criterion in arriving at an investment decision in each group of companies further, Kruskal-Wallis

Table 4.13 Descriptive statistics for the non-financial criteria

	Mean	Std. dev.	Median	Min.	Max.
Non-AMT companies					
Requirements of customers	4.17	0.94	4	2	5
Consistency with corporate strategy	4.00	1.07	4	1	5
Keeping up with competition	4.00	0.95	4	1	5
Quality and reliability of outputs	3.86	1.04	4	1	5
Obtaining greater manufacturing flexibility	3.52	0.93	4	1	5
The ability to expand in the future	3.50	0.80	4	2	5
Reduced lead-times	3.38	1.12	4	1	5
Reduced inventory levels	3.30	1.13	3	1	5
Improved company image	3.14	0.77	3	1	4
Employee morale	2.91	0.97	3	1	4
Experience with new technology	2.78	0.80	3	1	4
Less integrated companies					
Requirements of customers	4.13	0.87	4	1	5
Quality and reliability of outputs	4.13	0.71	4	3	5
Obtaining greater manufacturing flexibility	4.03	0.71	4	2	5
Consistency with corporate strategy	3.97	0.91	4	1	5
Reduced lead-times	3.93	1.01	4	1	5
Reduced inventory levels	3.78	1.18	4	1	5
Keeping up with competition	3.70	0.99	4	1	5
The ability to expand in the future	3.52	0.85	4	1	5
Improved company image	2.97	0.93	3	1	5
Employee morale	2.94	0.85	3	1	5
Experience with new technology	2.93	1.05	3	1	5
Fully integrated companies					
Quality and reliability of outputs	4.63	0.49	5	4	5
Requirements of customers	4.40	0.77	5	2	5
Obtaining greater manufacturing flexibility	4.36	0.74	5	3	5
Reduced lead-times	4.24	0.97	5	1	5
Consistency with corporate strategy	4.17	0.85	4	2	5
Reduced inventory levels	4.05	0.94	4	2	5
The ability to expand in the future	3.78	1.01	4	1	5
Keeping up with competition	3.63	1.26	4	1	5
Improved company image	2.95	1.15	3	1	5
Employee morale	2.84	0.86	3	1	5
Experience with new technology	2.72	1.09	3	1	5

Table 4.14 Results of Kruskal-Wallis test for non-financial criteria

	Mean Rank			Kruskal		
	Non AMT	Less integrate	Fully integrated	Wallis statistic	D.F.	P-value
Quality and reliability of outputs	36.82	41.19	59.32	15.0012	2	.0006
Requirements of customers	47.17	44.20	53.65	02.5874	2	.2743
Consistency with corporate strategy	47.55	44.74	50.64	00.9561	2	.6200
Reduced lead-times	32.33	45.53	54.46	10.8215	2	.0045
Obtaining greater manufacturing flexibility	31.26	44.45	55.17	13.1440	2	.0140
Reduced inventory levels	35.20	47.73	53.18	06.4487	2	.0398
Keeping up with competition	53.15	44.87	46.26	01.5043	2	.4713
The ability to expand in the future	42.77	43.89	52.77	03.1020	2	.2120
Improved company image	50.41	45.58	46.24	00.5187	2	.7716
Employee morale	47.68	47.10	44.13	00.3750	2	.8290
Experience with new technology	44.37	47.95	42.94	00.7098	2	.7012

one-way ANOVA was performed to test the importance of each criterion relative to the level of AMT.

Table 4.14 shows that there are significant differences between the three groups of companies for four non-financial criteria. These criteria are: 'quality and reliability of outputs' (p-value = .0006), 'reduced lead-times' (p-value = .0045), 'obtaining greater manufacturing flexibility' (p-value = .0140), and 'reduced inventory levels' (p-value = .0398). The rest have very low levels of significance. According to these results, it can be concluded that the importance of only these four non-financial criteria are affected by the level of AMT. They may thus represent major reasons for investing in AMT. The four criteria had increasing degrees of importance from non-AMT companies to less integrated companies, and from less integrated companies to fully integrated companies. Additionally, they had higher rankings across the three groups of companies in the same direction. These relationships are illustrated in table 4.15 which shows the mean importance for each criterion and its ranking within each group of companies.

Table 4.15 shows that the criterion 'quality and reliability of outputs' rises from being the fourth most important criterion in non-AMT companies to second in importance in less integrated companies and to the most important criterion in fully integrated companies. This indicates that 'getting

Table 4.15 The significant non-financial criteria in evaluating AMT investments

	Non AMT	Less integrated	Fully integrated
Quality and reliability of outputs			
Mean	3.86	4.17	4.63
Rank	4	2	1
Obtaining greater manufacturing flexibility			
Mean	3.52	4.03	4.36
Rank	5	3	3
Reduced lead-times			
Mean	3.38	3.93	4.24
Rank	7	5	4
Reduced inventory levels			
Mean	3.30	3.78	4.05
Rank	8	6	6

better quality and reliability of outputs' represents a major objective when investing in sophisticated manufacturing systems. The third most important non-financial criterion in both less integrated and fully integrated groups of companies is 'obtaining greater manufacturing flexibility'. The higher perceived importance of this criterion in AMT companies is consistent with the claims of suppliers of advanced technologies who often refer to flexibility as a major reason for acquiring sophisticated technologies. A number of writers argue that obtaining greater flexibility can benefit firms by increasing managerial control over output, reducing the risk of demand uncertainty and improving productivity (see for example, Tannous, 1996).

While the four non-financial criteria identified have significantly different degrees of importance across the three groups of companies, this is not true of virtually all the remaining criteria. It can be concluded that these criteria are seen as equally important or unimportant no matter what level of AMT investment is under consideration.

Risk Analysis of Investment Projects

The analysing of investment project risk represents an important element of the evaluation process (Butler et. al., 1993). The term 'risk' is used to refer to the uncertain nature of expected benefits, both tangible and intangible, from an investment project. In finance theory, it is generally accepted that as

Table 4.16 The importance of the risk analysis techniques (% of respondents)

	Score 1	Score 2	Score 3	Score 4	Score 5
Sensitivity analysis	2	8	13	31	46
Use of conservative cash flow forecasts	5	17	25	41	12
Adjust required payback period	15	15	23	24	16
Adjust required return on investment	18	18	31	21	11
Adjust discount rate	14	25	32	25	4
Probability analysis	40	27	21	10	2
Capital asset pricing model	53	22	13	10	2
Computer simulation	36	35	18	11	-

Score 1: Not important at all, Score 2: Not important, Score 3: Neutral, Score 4: Important, Score 5: Extremely important.

the risk of an investment increases so does the return that investors require to compensate for the extra risk (see for example: Drury et. al., 1993; Davis and Pointon, 1994; Lumby, 1995, Merrett and Sykes, 1980; Scott, 1990, and Reekie and Crook, 1995).

Risk analysis techniques are usually classified into two categories (see for example: Ho and Pike, 1991; 1992; and Walker and Klammer, 1984):

1. Simple (naive) techniques based on intuitive adjustments either to project cash flows or to a parameter in the evaluation model. For example, increasing the discount rate, reducing the required payback period, or using conservative cash flow forecasts.
2. Sophisticated techniques based on a comprehensive evaluation of uncertainties associated with the critical variables in the evaluation model. For example, sensitivity analysis, probability analysis, simulation, and the capital asset pricing model (beta analysis).

In order to investigate the role of risk analysis techniques, respondents were asked to indicate the importance of eight techniques using a five-point equal distance scale. Table 4.16 shows the percentages of respondents at each point on the scale. The eight risk analysis techniques can be seen to have very different levels of importance. The list is not mutually exclusive and most companies use more than one technique for analysing project risk. The table shows that the most important technique for analysing project risk is sensitivity analysis (77% of respondents rated it as either important or extremely important). This finding is consistent with earlier surveys both in the USA and UK (Klammer et. al., 1991 and Ho and Pike, 1991). Over 50% of respondents rated the 'use of conservative cash flow forecasts' as a means of handling a project risk as either important or extremely important. The three adjustment techniques: adjust required payback period, adjust required return on investment, and adjust discount rate, were rated as either important or extremely important by relatively few respondents (40%, 32%, and 29% respectively). The last three techniques in the table were considered to be either important or extremely important by very few respondents: probability analysis (12%), capital asset pricing model (12%) and computer simulation (11%).

Analysis of Risk Analysis Techniques According to the Level of AMT Table 4.17 shows descriptive statistics for the eight risk analysis techniques in the three groups of companies. Scrutiny of this table suggests relatively consistent variation in the importance of each technique in evaluating project risk across the three groups of companies. In all three groups, sensitivity analysis is the most important technique for assessing project risk. Except in less integrated companies, the minimum level of importance for sensitivity analysis was a score of '2'. This indicates the dominance of sensitivity analysis as a technique for assessing risk in practice. The 'use of conservative cash flow forecasts' is the second most important technique in all three groups of companies. Adjusting required payback period is relatively important in less integrated companies (mean score = 3.35) while it is on the middle point of the scale in both non-AMT companies and fully integrated companies. Adjusting the required return on investment and discount rate were not considered important. Finally, the table confirms that the least important techniques for all companies are probability analysis, the capital asset pricing model, and computer simulation. In order to test the impact of the level of AMT on the importance of each technique, Kruskal-Wallis one-way ANOVA was undertaken. Summary results of these tests are shown in table 4.18.

The results in table 4.18 show there is no significant difference between the three groups of companies at the 5% level for all of the eight risk analysis techniques. While, sensitivity analysis is significant at the 8% level, indicating reasonable statistical significance, the ordering of the mean values does not consistently rise or fall. These results suggest that although

Table 4.17 Descriptive statistics for the risk analysis techniques

	Mean	Std. dev.	Median	Min	Max
Non-AMT companies					
Sensitivity analysis	4.36	0.95	5	2	5
Use of conservative cash flow forecasts	3.50	1.05	4	1	5
Adjust required payback period	2.86	1.31	3	1	5
Adjust required return on investment	2.80	1.20	3	1	4
Adjust discount rate	2.67	1.06	3	1	4
Probability analysis	2.30	1.22	2	1	5
Capital asset pricing model	2.06	1.16	2	1	4
Computer simulation	2.05	0.97	2	1	4
Less integrated companies					
Sensitivity analysis	3.71	1.24	4	1	5
Adjust required payback period	3.35	1.19	3	1	5
Use of conservative cash flow forecasts	3.27	1.22	4	1	5
Adjust required return on investment	2.77	1.27	3	1	5
Adjust discount rate	2.70	1.22	3	1	5
Computer simulation	1.81	1.03	2	1	4
Probability analysis	1.65	0.88	1	1	4
Capital asset pricing model	1.50	0.76	1	1	3
Fully integrated companies					
Sensitivity analysis	4.24	0.89	4	2	5
Use of conservative cash flow forecasts	3.37	0.96	3	1	5
Adjust required payback period	3.13	1.31	3	1	5
Adjust required return on investment	3.03	1.32	3	1	5
Adjust discount rate	3.00	1.02	3	1	5
Computer simulation	2.19	0.98	2	1	4
Probability analysis	2.19	1.08	2	1	4
Capital asset pricing model	2.00	1.27	2	1	5

there are some variations in the ranking of the eight techniques in each group of companies, these differences are not significant.

Accola's Suggestion for Risk Measurement of Investment Projects Accola (1994) advocated the use of the dimension-oriented approach for considering project risk and suggested three dimensions for measuring the riskiness of

Table 4.18 Results of Kruskal-Wallis one-way ANOVA for the risk analysis techniques

	Mean Rank			Kruskal		
	Non AMT	Less integrated	Fully integrated	Wallis statistic	D.F.	P-value
Adjust required payback period	33.98	41.76	37.94	1.4792	2	.4773
Adjust required return on investment	35.47	33.75	38.07	0.5973	2	.7418
Adjust discount rate	34.40	34.74	39.52	1.0235	2	.5994
Use of conservative cash flow forecasts	41.63	37.00	37.72	0.6135	2	.7358
Sensitivity analysis	50.39	36.05	46.22	5.1265	2	.0771
Computer simulation	34.29	28.67	36.83	2.3758	2	.3049
Probability analysis	37.60	27.05	36.48	0.1320	2	.1320
Capital asset pricing model	33.67	25.95	32.05	2.5409	2	.2807

CIM investment decisions. The three dimensions and their associated measures are:

1. The Impact on Company Liquidity. This dimension could be measured using payback method. The project with the longer payback period is the riskier project.
2. The Variability of Project Outcomes. One measure that could reflect this dimension is the standard deviation of the expected return from an investment project. The larger the project's standard deviation the riskier the project.
3. The Possibility of Massive (Ruinous) Loss. The amount of capital investment as a percentage of the total assets of the firm might be used as an indication of the possibility of a ruinous loss. The larger the percentage of resources committed to a capital project, the greater the risk of insolvency.

For testing the applicability of the above three dimensions in measuring the riskiness of investment projects, respondents were asked to rate the importance of each dimension in arriving at investment decisions using a five-point equal distance scale. Table 4.19 gives the percentage of respondents for each point on the scale.

Table 4.19 The importance of the three dimensions of measuring project risks

	Score 1	Score 2	Score 3	Score 4	Score 5
Impact on company liquidity	10	14	21	34	21
Variability of project outcome	12	10	30	28	20
Possibility of massive loss	21	13	24	21	22

Score 1: Not important at all, Score 2: Not important, Score 3: Neutral, Score 4: Important, Score 5: Extremely important.

The table shows that the three dimensions were judged to be moderately important. Only the impact on company liquidity was rated as either important or extremely important by over 50% of respondents, while the other two dimensions, variability of project outcome and possibility of massive loss were rated 'important' or 'extremely important' by less than 50% of respondents. However, while the dimensions were not generally rated 'important' or 'very important' neither were they rated 'unimportant'. The three dimensions, impact on company liquidity, the variability of project return, and the possibility of massive loss, were rated as unimportant by only 24%, 22%, and 34% of respondents respectively. The importance of the three measures of project risk was further analysed according to the level of AMT in the company. Table 4.20 shows descriptive statistics for the three risk measures in the three groups of companies. Investigation of the results shown in this table reveals that there is little variation in the mean importance of the three measures of project risk. Only 'impact on company liquidity' scored a mean more than '3' in all three groups of companies. The other two measures: 'variability of project outcome' and 'possibility of massive loss' had mean scores of less than '3' in the less integrated companies.

To examine whether there were significant differences in the importance of each risk measure across the three groups of companies, Kruskal-Wallis one-way ANOVA was conducted. The results shown in table 4.21 reveal that there is a significant difference in the 'variability of project outcome' measure across the three groups of companies. This measure is less important in both less integrated and fully integrated companies than in non-AMT companies. It is also noted that 'possibility of massive loss' is significant at the 9% level and again the measure is less important in both less integrated and fully integrated companies than in non-AMT companies.

Table 4.20 Descriptive statistics for the three measures of project risk

	Mean	Std. dev.	Median	Min	Max
Non-AMT companies					
Impact on company liquidity	3.62	1.02	4	1	5
Variability of project outcome	3.70	1.13	4	1	5
Possibility of massive loss	3.68	1.29	4	1	5
Less integrated companies					
Impact on company liquidity	3.17	1.50	4	1	5
Variability of project outcome	2.65	1.37	3	1	5
Possibility of massive loss	2.70	1.58	3	1	5
Fully integrated companies					
Impact on company liquidity	3.48	1.19	3	1	5
Variability of project outcome	3.65	1.05	4	1	5
Possibility of massive loss	3.07	1.34	3	1	5

Table 4.21 Results of Kruskal-Wallis one-way ANOVA for Accola's risk measures

	Mean Rank			Kruskal		
	Non AMT	Less integrated	Fully integrated	Wallis statistic	D.F.	P-value
Impact on company liquidity	38.60	33.22	36.35	0.8072	2	.6679
Variability of project outcome	42.88	26.87	41.92	8.6996	2	.0129
Possibility of massive loss	44.68	30.85	35.65	4.8458	2	.0887

It can be concluded that only the 'impact on company liquidity' is moderately important in *all* companies -no matter the level of AMT the company employs. The other two measures are moderately important in non-AMT companies and fully integrated companies while their importance falls in the less integrated companies. The low importance of both 'variability of project outcome' and 'possibility of massive loss' measures in AMT companies may be attributed to the flexibility of these systems which, it is argued, increase managerial control over output, reducing the risk of demand uncertainty and improving productivity (see, for example, Tannous, 1996).

Table 4.22 The most frequently used payback period in the companies

	Non AMT (n = 20)	Less integrated (n = 30)	Fully integrated (n = 39)	All companies (n = 89)
Less than 1 year	-	3	3	2
1 - 3 years	70	73	80	75
4 - 5 years	25	23	15	20
More than 5 years	5	-	3	2

The Required Payback Period and Discount Rate Respondents, who use the payback method as a financial appraisal technique were asked to indicate the required payback period most frequently used in their firms. Table 4.22 shows the percentages of respondents for each group of companies. A period of 3 years or less is most frequently used by a very large percentage across all three groups of companies, 70% for the non-AMT companies, 76% for the less integrated companies, and 83% for the fully integrated companies. Higher percentages of respondents in both less integrated and fully integrated groups use shorter payback periods (3 years or less) than in the non-AMT group. A Pearson chi-square test was used to test whether these differences were significant.

In order to use the chi-square test all cells must include at least one observation (see, for example, Cramer, 1994, p. 83). Accordingly, data in table 4.22 cannot be manipulated statistically because there is no data in some cells. In order to use the test, the first two categories (less than 1 year and 1-3 years) were combined as 3 years or less and the last two categories (4-5 years and more than 5 years) were combined together as 4 years or more. The revised table and statistical comparisons are shown in table 4.23.

Results in table 4.23 show that there is no significant difference between the three groups of companies. Thus, it can be concluded that most companies in the three groups set a target of short payback targets and most AMT companies do not change this short payback target when evaluating AMT projects. Short payback is usually cited as a major obstacle to getting approval for AMT projects, which may have longer lives than conventional investments, and may have important cash flows in the later part of their lives (see, for example, Kaplan and Atkinson, 1989).

The questionnaire included a question asking those respondents who use one of the discounting methods to indicate the range of minimum rates of

Table 4.23 Results of Pearson chi-square test

	Non AMT (n = 20)	Less integrated (n = 30)	Fully integrated (n = 39)	Statistical comparisons
3 years or less	70	77	82	
4 years or more	30	23	18	
Pearson chi-square value				1.12135
P-value				.57082

Table 4.24 The minimum required rates of return or discount rates

	Non AMT (n = 17)	Less integrated (n = 23)	Fully integrated (n = 32)	All companies (n = 72)
Less than 10%	6	17	9	11
10 - 15%	53	30	31	36
16 - 20%	41	22	44	36
21- 25%	-	22	10	11
More than 25%	-	9	6	6

return or discount rates required by their firm, see table 4.24. The modal band for non-AMT companies and less integrated companies was in the range 10-15% (53% and 30% of respondents in each group of companies respectively), while it was in the range of 16-20% for fully integrated companies (44% of respondents). These results indicate an increase in the minimum discount rate from non-AMT companies to less integrated companies, and from less integrated companies to fully integrated companies. Of the non-AMT companies, 41% of respondents use a discount rate higher than 15%, this compares with 53% and 60% of respondents in less integrated and fully integrated companies respectively.

Table 4.24 was revised in order to undertake the Pearson chi-square test. The first two categories (less than 10% and 10-15%) were combined as 15% or less and the last three categories (16-20%, 21-25% and More than 25%) were combined together as 16% or more. Table 4.25 shows the result of this revision and statistical comparisons. There is no significant difference between the three groups of companies in the minimum required

Table 4.25 Results of Pearson chi-square test

	Non AMT (n = 17)	Less integrated (n = 23)	Fully integrated (n = 32)	Statistical comparisons
15% or less	59	47	40	
16% or more	41	53	60	
Pearson chi-square value				1.48024
P-value				.47706

rates of return or discount rates. However, it is evident that most less integrated companies (53% of respondents) and fully integrated companies (60% of respondents) use high hurdle rates (16% or more) and a large minority of non-AMT companies (41%) also use rates of 16% or more. Kaplan and Atkinson (1989) suggested that a rate between 12% and 13% or between 8% and 9% (net of inflation) is the most appropriate rate in discounting cash flows.

Discussion

Examination of the survey results indicates that investment projects are affected by three groups of factors: financial return measures, non-financial criteria, and risk measures. For example, the payback period, a financial measure, was judged important or extremely important by 75% of respondents. Eighty seven percent of respondents considered the 'quality and reliability of outputs', a non-financial criterion, important or extremely important. 'Sensitivity analysis', a risk measure, was considered important or extremely important by 77% of respondents. The following discussion of the survey results is based on these three groups of factors.

Financial Return Measures

The survey results indicate that, while all measures of financial return are important, the naive financial appraisal techniques (ROI and payback period) are seen as marginally more important than the sophisticated financial return measures (NPV and IRR) in all three groups of companies (non-AMT, less integrated, and fully integrated companies). The difference in the importance of each technique between the three groups was tested using the Kruskal-Wallis one-way ANOVA and no significant differences were revealed. Therefore, it seems that the sophistication of the project being evaluated does not make a significant difference to the importance of the financial appraisal technique(s) used.

Recognising that most companies use more than one financial appraisal technique (as shown by the current and almost all previous capital budgeting surveys), it would be useful to know the perceived importance of each technique. According to the results of the current survey (tables 4.8 and 4.10), financial appraisal techniques can be classified into three levels of importance: first, the most important level includes payback period and ROI, second, the second most important level includes IRR and NPV, and third, the least important level: discounted payback period.

The role of ROI as one of the most important techniques is quite surprising. Previous surveys indicated a decline in the use of ROI (see for example: Klammer et al., 1991; Kim et al., 1986; Chen, 1995; and Kalyebara, 1996). Lumby (1995, p. 49) gives two reasons for expecting the popularity of ROI to decline, both reasons arising from the high rates of inflation suffered by the UK economy in the late 1970s and 1980s. First, the high rates of inflation led to high interest rates and this high time value of money brought home to management the importance of taking the time value of money into account in decision making. (With interest rates of around 20% it becomes obvious that the timing of cash flows is important.) Second, the high rates of inflation and the failure of the inflation accounting debate taught management about the doubtful validity of accounting profit. Thus a method that uses accounting profit and ignores the time value of money has little to recommend it as a decision making tool.

There is, however, conflicting evidence. In Pike's study (1988a), it was found that 56% of respondents in 1986 were using ROI in their evaluation of investment projects, while the percentages of ROI users in 1981 and 1975 were 49% and 51% respectively. The wide acceptance of the use of ROI as a financial appraisal technique for investment projects, as found by Pike and confirmed in this survey, indicates managers' preference for relating the project's return to the return on the business as a whole (see for example: Pike, 1982; Kee and Bublitz, 1988; and Lumby, 1995, p. 47). This might be consistent with the results of Lefley's study (1994a) in which he found that almost 46% of respondents were using ROI as a management appraisal method. Pike and Wolfe (1988) explained this preference for ROI as an investment appraisal technique by practitioners as follows:

> '... its (ROI's) absence leads to an inconsistency between methods used to report a firm's operating results and the techniques most frequently used to appraise investment decisions. This inconsistency is most actually experienced where the manager of an investment centre is expected to use a DCF approach in reaching his investment decision while his short-term performance is being judged on a return on investment basis. Little wonder then that the divisional manager generally shows a marked reluctance to enter into those profitable long-term investment decisions producing lower returns in early years'. (pp. 22-23)

Accola (1994) also claims that using ROI as a financial measure for evaluating investment in new technology is justifiable because of the lack of 'enough reliable information about future cash flows ... to compute discounted cash flow measures'.

Using ROI in appraising investment projects has been criticised from a theoretical point of view for five reasons (see, for example, Hirst, 1988; Reekie and Crook, 1995; Lumby, 1995; and Samuels and Wilkes, 1986). First, the ROI is an ambiguous concept because there is no generally accepted method for its calculation. The method might be abused by allowing the decision maker to select a definition of ROI which best suits his/her preconception of a project's desirability. Second, ROI is based on the concept of accounting profit rather than cash flow. It is argued that use of accounting profit, a 'reporting' concept, is entirely incorrect in a decision making context. As Lumby (1995, p. 49) pointed out, capital investment is an economic or resource allocation decision, and the economic unit of account is cash. It is cash, not accounting profit, which gives power over resources. Third, ROI does not consider the differing lives of investment projects. For example, consider two investment projects, the first project gives ROI of 20% for 10 years while the second project gives ROI of 21% but for only three years; the first project might be the more desirable for shareholders, but ROI would favour the second project (Reekie and Crook, 1995, p. 378). Fourth, ROI ignores the time value of money. Finally, it is argued that ROI does not link well into long-term strategy and more 'strategic' considerations such as market advantages (Finnie, 1988).

The preference for the payback technique for evaluating investment projects confirms the earlier findings of Klammer (1970), Woods et al.

(1985), Schall et al. (1978), Pike (1982 and 1988b), Drury et al. (1993), and Lefley (1994a). For example, in Drury et al's (1993) study, the payback method was ranked most important with average score of 1.94 (scale of 1-5 where 1 = most important and 5 = least important). Similarly, Lefley (1994a) found that the payback method was ranked first by the highest number of responding companies (32% of his respondents). The preference for the payback method, however, is contrary to the earlier findings of Hendricks (1988), Kim et al. (1986), and Chen (1995). For example, Hendricks (1988) reported that 29% of respondents (in a sample of 85 companies) were using payback as a primary financial appraisal technique while 55% were using the IRR. Also, Chen (1995) reported that the mean importance of the payback method (scale of 1-5 where 1 = not important and 5 = extremely important) was 3.096, 3.235, and 3.165 for three types of project: replacement, expansion existing-products, and expansion new-products respectively, while the importance of DCF methods was 3.678, 4.226, and 4.252 for the three types, respectively.

The payback period as a criterion for appraising investment projects has been subjected to much debate, and prima facie should not be used as a decision making tool. For example, Samuels and Wilkes (1986, p. 324) argue that: '... payback period methods should be matters for financial reporting only ...'. Hirst (1988, p. 29) indicates that '... payback is ... a very poor way of making investment decisions ...', and Lumby (1995, p. 46), also, argues that '... it (the payback) should not really be used to give decision *advice* at all, but only to give *information* on the speed of return of the initial outlay, which may or may not be of relevance ... in the decision processes'. Two of the most common objections against the payback period as an investment decision making method are that, first, it ignores the order of cash flows within and all cash flows beyond the payback period, and second, it does not consider the time value of money (see for example: Lumby, 1995; Hirst, 1988; Randhawa and West, 1992; and Merrett and Sykes, 1980). It might be argued, however, that these two common objections offset each other because the error from ignoring the post payback cash flows may be compensated by the error arising from failure to account for the time value of money. Accordingly, Pike (1988a, p. 25) concluded that, for ranking projects, the payback period method provides a reasonable approximation to NPV based decisions provided that the early cash flows are a reasonable predictor of later cash flows. In fact this conclusion is suspect because it assumes that the expected error from ignoring the time value of money can be compensated by the error from ignoring the post payback period cash flows, regardless of the length and size of the mutually exclusive projects under consideration.

In addition to these objections, Lumby (1995, p. 44) drew attention to the ambiguity of the payback period as a decision rule. This ambiguity relates to two aspects of the payback calculation: first, the definition of the outlay that should be recovered when the project being evaluated includes a negative cash flow during its operating life, and second, the definition of the start of the payback time period. The latter problem can arise if the project under consideration has an extended implementation period: is the payback time period to be counted from the commencement of the project or from the completion of project outlays? If there is ambiguity in interpretation of any technique designed as a decision aid, Lumby (1995, p. 44) warns:

> ' ... it (the technique) is likely to be manipulated so as to lend backing for the desired decision, rather than the right decision. Any decision rule open to such misuse is dangerous and must be viewed with suspicion'.

It is also argued that the lack of academic support for payback is due to the fact that the method might not be consistent with the objective of maximising company owners' wealth (Lefley, 1994a). Only discounted cash flow methods are truly consistent with this objective (see for example: Lumby, 1995 and Clark et al., 1984). Nevertheless, it can be argued that there is a relationship between the payback period and the IRR (see for example: Davis and Pointon, 1994, p. 47; Pike and Dobbins, 1986, p. 284; and Weingartner, 1969). The reciprocal of the payback period is the IRR of a project of infinite life (or a good approximation to this rate for a long-lived project) provided that the project generates a relatively constant annual cash flow. For example, a four-year payback period could be converted to a 25 per cent IRR[2]. However, regardless of the relationship between the payback period and the IRR, it is widely accepted that the only discounting method which could properly achieve the objective of maximising company owners' wealth is NPV not IRR. This is mainly because of the re-investment assumption implicit in the IRR calculation (see, for example, chapter 6 in Clark et al., 1984 and Lefley, 1997).

Despite the above argument against the use of the payback as an investment appraisal tool, its widespread use must indicate that it has some merit. The main advantages usually cited in the literature include its simplicity of calculation and understanding. Further, many authors have tried to investigate why practitioners prefer the use of the payback method over more theoretically sound methods such as NPV and IRR (see, for example, Weingartner, 1969; Kee and Bublitz, 1988; Pike, 1985; Pike and Dobbins, 1986; Booth, 1996; Grinyer and Green, 1997). Investigations have resulted in three reasons for the widespread use of the payback method in practice. They can be summarised (Pike and Dobbins, 1986, pp. 383-386):

1. The use of payback period as a profitability measure. This is due to the relationship between payback period and IRR discussed earlier. Remember that this is subject to two assumptions: the project's life must be infinite or at least long, and the annual cash flows generated by the project must be stable.
2. The use of the payback period as a constraint. It is argued that the payback period can act as a simple, reasonably efficient measure for ranking projects when constraints prevail. This may be done in two ways: first, as a first-level screening device so that only a handful of the investment ideas originally generated need be subjected to more thorough financial investigation. Second, the payback period may serve as a device for ranking projects according to their ability to repay quickly if the firm is experiencing a liquidity problem.
3. The use of the payback period as a tool reflects the project's risk characteristics. This method assumes that risk is time-related on the basis that the longer the period, the greater the chance of failure. Accordingly, a project with shorter payback period is less risky than one with a longer payback period.

Of course using the payback method gives priority to short-term projects, and these may be less profitable than long-term projects. This problem is especially serious in the case of AMT projects. These projects, as discussed in the first two chapters, tend to need long-term appraisal because their full benefits often cannot be realised within a short time span. In this respect, Primrose (1991, p. 22) argues that 'payback should never be used for investment appraisal, because its use will result in companies making the wrong investment decisions ...', also, Lefley (1994a) points out that '... the use of the payback method of investment appraisal practically guarantees rejection of AMT projects ...' (p. 2758) and '... the payback method, with its 'short' payback period, is inappropriate for projects such as AMT where the capital expenditure is spread over a number of years, and the returns do not fully materialise until well into the life [of] the project'.

The low importance of the sophisticated financial appraisal methods (NPV and IRR) as compared to the naive methods (payback and ROI) is consistent with the findings of Woods et al. (1985) and Lefley (1994a). For example, in Lefley's study, 55% and 52% of respondents used IRR and NPV as one of their financial appraisal techniques for AMT projects, while, 68% of respondents used the payback method. However, the results are inconsistent with those reported by Klammer and Walker (1984), Kim et al. (1986), Hendricks (1988), Chen (1995), and Kalyebara (1996), who found that the percentages of companies using sophisticated financial techniques were higher than for those using the payback method.

Importantly, the results of this survey are consistent with those of Woods et al. (1985) and Lefley (1994a), who also concentrated on AMT investment projects. It might be tempting to suggest an inverse relationship between the sophistication of an investment project and the sophistication of the technique used in its evaluation. However, caution should be exercised in reaching such a conclusion: first, the current survey suggests that *all* the measures of financial return are important in investment evaluation and, second, the Kruskal-Wallis one way ANOVA test revealed no significant differences between companies employing different levels of AMT.

It was expected that the sophisticated methods (NPV and IRR) would receive the highest importance in the evaluation of investment projects. This expectation was based on three points. First, companies surveyed in this study were the largest in the UK[3]. Second, the survey was aimed primarily at AMT project appraisals and according to Woods et al. (1985, p. 42), '... users (of new technology), by definition, are more advanced in their thinking and therefore more likely to use more modern investment appraisal techniques with their greater sophistication.'. In addition, Kim et al. (1986, p. 50) stated that: '... Clearly, discounted cash flow methods are preferred when evaluating major investment projects'. Finally, widespread availability of user-friendly computer packages allows the calculation of either NPV or IRR in a few seconds. However, three possible reasons might explain why the expectation was not supported. First, a fundamental assumption underpinning the theoretical preference for DCF methods is that the primary objective of the firm is the maximisation of shareholders' wealth. The relationship between the wealth maximisation objective and the use of DCF method(s) was examined by Chen (1995) who found a positive correlation between the objective of maximising shareholders' wealth and DCF methods. Nevertheless, Pike and Dobbins (1986) found that the most common investment objective is return-on-assets with shareholders' wealth maximisation having a relatively low priority. Second, there is a little empirical support for the association between investment sophistication and better performance (see for example: Klammer, 1973). Finally, as shown in the current survey, non-financial criteria receive a high importance in the investment decision process, so it may not be surprising to find a company relying heavily on the judgement of decision maker(s) supplemented by simple financial methods. As Pike and Dobbins (1986, p. 276) note: '... formal evaluation such as DCF is just one of many factors in the investment decision process'.

Perhaps the most surprising feature of the results is the stringent payback and discounting criteria employed by most companies. There was no significant difference between companies in the three groups, most employed short payback periods (3 years or less) and high discount rates (16% or more). Given that the benefits of AMT are of a long term nature, one might have expected companies investing using AMT to employ less stringent criteria. As noted earlier, many authors warn that such stringent criteria lead to rejection of AMT projects (see for example, Lefley, 1994a; and Kaplan and Atkinson, 1989). Two comments: first, companies investing in AMT might demand high financial returns because of the very large capital sums involved, and second, it is not clear how much the apparently high financial returns demanded are offset when AMT companies also take non-financial factors into account.

To summarise, it is clear that financial analysis is very important in investment decision making. Most respondents disagree with the proposition that: 'it is possible to implement an AMT investment proposal based only on potential non-financial benefits'. Also, most companies employ a range of financial indicators and consider most of these to be important with simple measures being slightly preferred over sophisticated ones.

Regardless of the financial appraisal method(s) used in evaluating investment projects, the survey indicated that some of the benefits of investing in AMT may be overlooked because they cannot be subjected to financial evaluation. Most respondents believed that 'some of the potential benefits of AMT are not taken into account because they are difficult to express in financial terms'. Obtaining experience of new technology, effects on employee morale, and improved company image are examples of benefits that many companies did not consider in the evaluation process either financially or non-financially. However, respondents indicated that while a number of AMT benefits may not be included in financial evaluation, they are considered qualitatively in the evaluation process. Examples of such qualitative aspects include the quality of the output and obtaining greater flexibility. It seems practitioners would not support Primrose (1991) who claimed that all the benefits of AMT can and should be expressed in financial terms. The majority of respondents included the following benefits in their financial evaluation: reduced labour costs, reduced material costs, reduced inventory levels, reduced scrap and rework costs, increased sales volume, savings from less frequent set-ups and increased manufacturing capacity.

Non-financial Criteria

The survey results indicated that at least eight non-financial criteria (or intangible factors) were considered important in the evaluation process. They were: 'requirements of customers', 'consistency with corporate strategy', 'keeping up with competition', 'quality and reliability of outputs', 'obtaining greater manufacturing flexibility', 'the ability to expand in the future', 'reduced lead times', and 'reduced inventory levels'. A separate question (see table 4.4) confirmed that, except for reduced inventory levels, these factors were usually evaluated non-financially for AMT investments. Thus, although reducing inventories is an important criterion when evaluating investment projects, it is usually reflected in the financial appraisal. This confirms the view of Kaplan and Atkinson (1989, p. 485) that 'reductions in inventory levels represent a large cash inflow at the time when the new technology becomes operational and the inventory can be reduced'.

Further analysis regarding the importance of intangible factors across the three groups of companies indicated that these factors were seen as more important by fully integrated companies than by non-AMT and less integrated companies. Evidence of this view is provided by the higher levels of both the means and the minimum scores of most factors for fully integrated companies than for the other two groups of companies (see table 4.12). The higher importance of non-financial factors supports the view of Meredith and Hill (1987) who argued that the justification of higher levels of automation usually requires more attention to intangible factors relative to economic justification. For four non-financial factors the Kruskal-Wallis one-way ANOVA indicated *significant* differences among the three groups of companies. These factors were 'quality and reliability of outputs', 'obtaining greater manufacturing flexibility', 'reduced lead times', and 'reduced inventory levels'. The survey provides evidence that these factors take on increased importance when evaluating AMT investments and may be the key 'drivers' behind such investments.

The non-financial factor 'requirements of customers' was most important in both non-AMT and less integrated groups of companies and the second most important factor in the fully integrated group of companies. This may indicate that firms should pay considerable attention to market research during investment appraisal.

The importance of 'consistency with corporate strategy' declined from being the second most important criterion in the non-AMT group to fourth and fifth in less integrated and fully integrated companies respectively. Perhaps this is because the issue is not stressed in the detailed evaluation of the most sophisticated systems since the consistency of such projects with corporate strategy should be investigated very early, before proceeding to detailed evaluation of the project. This view was expressed by most companies in the field study reported in the next chapter.

It was surprising that the criterion 'employee morale' was judged unimportant in all three groups of companies. Many writers consider this factor to be one of the most basic motivations for investing in AMT (see for example, Accola, 1994; and Mensah and Miranti, 1989). However, this finding confirms the earlier finding of Pike et al. (1989) that employee morale is one of the least important factors considered when evaluating investment projects. They also commented:

> 'This is perhaps a cause for concern as far as manufacturing industries are concerned, since in our interviews the problems of shortage of skilled manpower and the difficulty of retaining them have been mentioned by a number of such companies, and several have observed that it is increasingly important to their workforce to know that they have the opportunity to work with the best technology available'. (p. 19)

The importance of non-financial factors in investment appraisal was thus confirmed by this survey, and specific factors are identified as important by the majority of respondents. The four factors identified as key driving forces behind AMT decision making should receive particular attention in the decision making process.

Risk Measures

The survey results indicate that sensitivity analysis is the most important technique for dealing with risk in most of the sample companies. This is consistent with the earlier findings of Klammer et al. (1991), Pike (1988b) and Ho and Pike (1991). In addition, a significant difference[4] was found regarding the importance of sensitivity analysis between the three groups of companies. It was less important in both less integrated and fully integrated companies than in non-AMT companies. This finding is consistent with that of Lefley (1994a) who also found a reduction in the use of sensitivity analysis in assessing AMT investment risk.

The widespread use of sensitivity analysis may be attributed both to its simplicity and to the availability of computer packages which can help apply it in practice. In fact, this technique cannot be used to quantify risks, it merely describes the sensitivity of key variables through their effect on the profitability of a project. It gives an indication of the sensitivity of the most important variables, but not the probability of their occurrence. A further limitation of sensitivity analysis is the consideration of only one variable at once. In other words, it is assumed that all variables remain constant except the one under consideration. This assumption is obviously unrealistic. It can be overcome by using the computer simulation technique. However, this method was judged important by only a few respondents (11%). This does not mean that sensitivity analysis is misleading but it does mean that it should only be a first step in evaluating risk. A knowledge of the most sensitive factors enables decision makers to explore their variability and to undertake more sophisticated methods of risk analysis.

Adjusting the required payback period according to riskiness of a project was judged as either important or extremely important technique by only 40% of respondents. This is not consistent with the earlier finding of Lefley (1994a) who found 71.5% of respondents used this method. However, Lefley's respondents may have used the technique without considering it especially important, and when this is taken into account, the results of the two surveys can be reconciled. The low percentage of respondents who considered adjusting the required payback period as an important way of reflecting the project risk may seem inconsistent with the high importance given to payback as a financial appraisal technique. However, it provides evidence that decision makers separate the two issues of financial evaluation and risk assessment.

In this survey, 70% of respondents in non-AMT companies required a payback of 3 years or less while 76% and 83% of respondents in less integrated and fully integrated companies, respectively, required the same payback period. If shorter payback is required for higher risk investments, it might be tempting to infer that AMT investments are perceived to be riskier than other investments. A similar inference might be drawn from the finding that higher discount or hurdle rates are employed in both less integrated and fully integrated companies than in non-AMT companies. While only 41% of respondents used a hurdle rate of 16% or more in non-AMT companies, 53% and 60% of respondents used the same range in less integrated and fully integrated companies respectively. However, these inferences are not supported because the results of Pearson chi-square tests show no significant differences between companies in either the required payback period or discount rate. The results of these tests suggest that AMT companies employ similar criteria to these employed by non-AMT companies.

Generally, the survey findings relating to risk analysis did not give a clear indication of how practitioners deal with investment risk. Only sensitivity analysis and conservative cash flow estimates were considered really important in dealing with risk by the majority of respondents. Although large minorities considered the adjustment of required payback/ROI/discount rate as important, the most preferred theoretical methods, probability analysis, the capital asset pricing model and computer simulation, were judged important by only very few respondents. Furthermore, when respondents were asked to indicate the importance of the three possible dimensions of risk suggested by Accola (1994): impact on company liquidity, variability of the project outcome, and possibility of massive loss, only the impact on company liquidity was rated as important by over 50% of respondents. These results led to more detailed investigations of risk measurement during the field study discussed in the next chapter.

Conclusions

The survey suggests that relatively few UK companies have invested in fully integrated manufacturing systems. Despite the sample being drawn from the largest manufacturing companies in the UK, only 43% of responding companies had invested in fully integrated manufacturing systems such as FMS and CIM. However, a large percentage of these investments had been recently implemented and growth in investment in the most sophisticated manufacturing systems, such as CIM, might be expected in the next few years. This expected growth is also supported by a recent study by Hamblin and Lettman (1996) who found a bicausality between firm profitability and AMT investment.

Consistent with the literature, the survey results indicated that the most difficult task in the evaluation of AMT projects was that of quantifying intangible factors. It was recognised by respondents that many AMT benefits may be overlooked because they are difficult to express in financial terms. Further, it was agreed that the non-financial benefits of AMT per se were insufficient to gain approval for AMT projects. One could conclude that financial criteria dominate investment decisions in AMT, and indeed, other investment decisions too. However, a large minority of respondents (48%) did not agree that financial criteria represent an obstacle in the approval of AMT projects.

Regarding financial measures in the appraisal of investment projects, the survey results reported here confirm previous empirical results that practitioners give the highest importance to the non-discounting techniques (payback and ROI). In addition, the IRR technique was rated higher than the theoretically favoured method, NPV. The Kruskal-Wallis one-way ANOVA did not indicate significant difference in the importance of any financial technique whether evaluating AMT or any other investment. This suggests that the sophistication of the project being evaluated does not lead to an increase in the sophistication of the financial evaluation techniques used. In other words, the nature of the project being evaluated is not a factor in determining the importance of any particular financial appraisal technique. It was also confirmed that firms usually use several techniques when evaluating investment projects although they may consider them to have different levels of importance.

It is often argued that traditional approaches based on financial measures, whether discounting or non-discounting methods, penalise long term projects such as AMT through high discount rates, or required ROI/IRR, and short payback targets, and by failing to capture many of the intangible benefits which should flow from AMT investments. The results of this survey would appear to support this view. Firstly, a period of 3 years or less was set as a target for payback by most of the sample companies. This seems very short, especially for AMT projects where benefits often materialise in the later years, (the survey shows that most companies expect a useful life of at least 7 years for most types of AMT project). Secondly, the majority of AMT companies used a high discount or hurdle rate, typically, 16% or more. Finally, the survey results indicate that many intangible benefits may not be considered in the evaluation process for AMT project because they are not easily quantifiable. This was confirmed in two ways. First, by the agreement of the majority of the respondents with the statement that: 'some of the potential benefits of AMT are not taken into account because they are difficult to express in financial terms'. Second, only 7 benefits out of 21 benefits listed in the survey were considered financially in the evaluation of AMT projects by the majority of the respondents. Nevertheless caution should be exercised in interpreting these findings because respondents did not generally agree with the proposition that: 'it is difficult to get AMT investment proposals approved because of stringent financial criteria', and that many 'intangible' benefits are considered even though not included in financial analysis.

The apparent contradictions in the responses may, to some extent, be resolved when one considers that decision makers make judgements on the basis of a number of factors, including those which cannot be easily quantified. This survey provides convincing evidence that such factors are important in AMT decisions. In the companies with fully integrated AMT systems a number of 'intangible' factors were rated much more importantly than in the other companies. There were statistically significant differences between the different groups of companies in respect of 'quality and reliability of outputs', 'obtaining greater manufacturing flexibility', 'reduced lead-times' and 'reduced inventory levels'. These factors were seen as increasingly important from non-AMT to less integrated to fully integrated AMT companies, and it is concluded that these factors are often the driving forces behind AMT investment decisions. Further, four non-financial factors were rated highly by most respondents in the three groups of companies: 'requirements of customers', 'consistency with corporate strategy', 'the ability to expand in the future', and 'keeping-up with competition'. Thus, these four factors can be considered important for investment decisions in general (whether AMT investments or not). However, it was not clear how these factors are used in the evaluation process for investment projects. This led to more exploration in the field study which will be discussed in the next chapter.

Surprisingly, a number of other factors that are often quoted in the literature as key benefits of AMT systems were not rated especially important by respondents. These factors included: 'improved company image', 'employee morale' and 'experience with new technology'.

Respondents indicated that project risk is difficult to assess and that, despite the attention the topic has received over the last three decades, practitioners still apply very simple techniques. Sophisticated techniques such as probability analysis, simulation, and the capital asset pricing model were the lowest rated techniques in analysis of the riskiness of investment projects. Only sensitivity analysis and the use of conservative cash flows forecasts were judged important by the majority of respondents. The simplicity and availability of computer packages to perform sensitivity analysis may explain the widespread use of this technique in the analysis of risk. Although the majority of respondents indicated that they use short payback periods and high hurdle rates, they do not consider adjustments to these to be an important way to consider the risks of an investment project. Further, the three dimensions of risk suggested by Accola (1994) were judged by most of respondents to be only moderately important. The survey results regarding risk did not give a sufficiently clear idea about how practitioners deal with the risk of investment projects. This issue was further investigated in the field study reported in the next chapter.

Notes

1 Lefley's study was conducted in 1992.
2 It should be noted that if the project's life is short, the reciprocal of the payback period over-estimates the true IRR.
3 Most previous surveys have found that there is a significant relationship between the size of the company and the sophistication of the method(s) used in evaluating investment projects. The largest companies usually use the most sophisticated method(s) (see for example: Drury et al., 1993; and Chen, 1995).
4 At 8% level of significance.

5 The Field Study of AMT Investment Decisions in Some Large UK Companies

The survey results reported in the previous chapter indicate that both financial return and intangible benefits are important factors in the evaluation of AMT investment projects. However, a substantial majority of respondents indicated that AMT investments could not be justified on non-financial grounds alone. Additionally, when asked if specified financial measures were important in evaluating AMT proposals, the majority of respondents indicated that measures such as ROI, IRR, NPV and payback were important. In relation to the intangible benefits of AMT, most of the respondents felt that the impact of AMT investment on product quality, customer requirements, manufacturing flexibility and overall strategy were either 'important' or 'extremely important'. A separate question confirmed that such intangible factors were usually evaluated using non-financial criteria.

Insofar as risk is concerned, the findings of the survey are more equivocal. In the assessment of risk, the majority of respondents considered only sensitivity analysis to be really important. At best there was lukewarm support for the adjustment of cash flows or target returns/payback to take account of risk, and techniques such as probability analysis and simulation were little favoured. In respect of the possible aspects of risk identified by Accola (1994): variability of project cash flows, impact on liquidity and possibility of ruinous loss, respondents again did not provide clear evidence. Whilst some respondents did rate these issues 'important' or 'extremely important', a slightly smaller number regarded them as relatively unimportant.

Some aspects of AMT investment decisions cannot easily be investigated using a questionnaire, and so in order to pursue these aspects and to provide deeper explanations of some of the results of the survey, a field study was undertaken. The following research questions were investigated:

1. What are the main motivations for investing in AMT?
2. What are the key factors affecting the decision to invest in AMT (financial, non-financial, and risk factors)?
3. Why are specific financial return measures preferred over the other measures?
4. If non-financial factors are considered, how are these factors reflected in the decision model?
5. What sorts of risk are encountered in AMT investment? If any:
6. How can these risks be measured and incorporated in the investment decision model?
7. What is the process of evaluating investment projects?

Nine in-depth interviews with respondents to the questionnaire who had experience in evaluating AMT projects were conducted, and the transcripts of each interview together with publicly available information were used in producing a series of case studies[1]. The primary aim of each case was to illustrate how the investment decision for a specific project was undertaken within the company. In addition the case studies highlight the main reason(s) for the investment, the factors affecting the decision and the process of the decision making, and the formal presentation of the information relating to the investment. Each case starts with a brief company background[2] and a description of the project under study, then specific information and decision making processes are considered.

Seven cases are presented in the following sections[3]. They are followed by some discussion and conclusions.

Case 1 - Baby Food Ltd

Background

Baby Food Ltd is a private company, wholly owned by a Dutch parent. It produces about 4 million baby meals in glass jars per year and had a turnover for the year ending 1994 of £99,129,000. There are about 200 people on the site concerned, over half of them 'direct' factory operators and the rest support staff: planning, finance, systems, engineering etc. The company is part of the food processing industry with major capital investment in automated equipment. Fresh food is received, converted into baby food and packed in glass jars ready for sale; the whole process taking about two hours.

Food preparation is the most labour intensive part of the process as meat, vegetables and fruit are prepared prior to entry to the automated production lines. Two production areas (for the preparation of wet and dry food respectively) feed two automated production lines (a sterilised line and a pasteurised line), and some 130 different products are produced. The sequence in which products are produced can affect the time taken to change-over from one product to another, and optimisation of both sequence and length of production runs is important in the plant.

Paper Labeller Project

Packaging the product is an important part of the process, and the interviewee was prepared to discuss a recent investment decision to switch from printed plastic sleeves to paper sleeves on the glass jars. This involved the installation of new automated equipment costing between £160,000 - 170,000.

The plant provides output to sister operations at a transfer price, none of the plant's output is sold direct to retailers, and the plant's aim was specified as providing high quality products at the lowest possible prices. The main justification for the investment was the expectation that operating costs would be reduced thus helping the plant to lower its transfer prices.

The major technical advantage in switching from plastic to paper sleeves is in printing, it is much easier to print onto paper than onto plastic. A high quality paper labeller was needed so that break-downs would be minimised, and uninterrupted operation of the equipment would be possible at a rate of 500 jars or more per minute.

Factors Affecting the Decision

The main driver for investing in this new equipment was purely the projected cost savings. The use of paper labels instead of plastic leads to a reduction in the cost of printing. Also, paper labels are more environmentally friendly than plastic labels.

The parent company sets guidelines for capital expenditure. There is a budget for what is called 'maintenance expenditure', but projects like the paper labeller are 'special projects'. Such projects have to be justified in their own right and in more detail. The paper labeller project would have an impact on both the factory and the market place.

> '... you're changing your format and so it has an impact across a whole business territory ... cost of labels, possible write-offs of old stocks ... you also have to make sure that you can sell that format ... we then have to justify that in terms payback ... we also do internal rate of return'.

The payback period is a very important measure. It is almost like a screening method because it is used as a first indicator before continuing the analysis to calculate internal rate of return. Net present value was calculated but only as a 'by-product' of IRR. The directors prefer IRR because it is more understandable. A minimum IRR of 22% is expected if a project is to be accepted. This rate is set by the parent company and used for all capital projects.

It was, of course, important to be sure that the change would be acceptable in the market place, and so there was close co-operation with individual marketing units in putting together the case for the change. Sensitivity analysis was also undertaken.

> '... If you had some disaster like paper prices going through the roof, like they did last year in fact,... then you are reducing that central return. So we did a lot of sensitivity around 'what if' the prices went this way or 'what if' the prices went that way, but we didn't actually change anything about the 22%'.

Therefore sensitivity analysis was undertaken but no attempt was made to adjust the required discount rate to allow for a 'risky' project.

Decision Making

Question: Suppose you didn't get 22%, for example, suppose you got 18%?

> 'It wouldn't be totally ruled out. My belief is, if you got it above 16 - 17% and showed that it was a good idea and that there were some other non-financial benefits, then you are in with quite a fighting chance'.

The interviewee went on to say that the group was becoming increasingly stringent in demanding good financial returns on all projects.

Question: What weighting might be given to financial return versus intangible benefits?

> 'I would give 70 for return and 30 for benefits.'

Question: Where does the risk fit into that?

> 'You would quantify that in your sensitivity analysis. You would have to make a judgement then'.

The interviewee agreed that the decision would be based on a synthesis of judgements concerning financial return, intangible benefits and risk. He noted that these factors could vary significantly between projects. For example:

> '... There are times when we might make other investment decisions which won't necessarily show a 22% internal rate of return, but you do that because quality matters. We need to produce a quality, safe, product. If you don't do that then you have no market place. So there are times when you invest in things like an X-ray. Perhaps it won't magically increase your sales or anything. It won't reduce your costs, it will do the opposite. You will do that because the market actually demands a measure of rigorousness if you like, in the manufacturing process.'

Formal Process and Presentation

The evaluation process is very formal in the company. A standard presentation format is adopted for all investment proposals. Summary of recommendations, benefits, risks and costs are the main headings in the proposal.

> '... We try and keep the same format because everyone expects to see a certain format and it makes explaining it more straightforward rather than relearning each time'.

The presentation includes 'a one page summary of costs' together with narratives documenting benefits and risks. No attempt is made to quantify or 'score' all the benefits or risks in a formal manner. Strategic issues are considered first, then the formal analysis and project justification are set out.

> '... the factory here will put together the capital application form and we will send that through. It won't be done by the marketing groups [the companies that sell the product], obviously they will be supporting us in terms of when their group directors come back and say what do you think of this idea. They are already on board. We will fill in the capital form with all justification behind it and send that through the main channels. That's [to] the head office and for that case [the paper labeller] to Holland as well'.

The proposal is signed off by finance and by the local site manager. The managing director may then have authority to sign off the project, but this depends on the proposed capital expenditure, and the paper labeller project needed to be authorised at a higher level.

The investment decision in the company is summarised in figure 5.1.

Case 2 - Engineering Plc

Background

Engineering Plc is a very large company consisting of about 55 subsidiary companies in various industries. The group activities are automotive engineering, defence and aerospace, marine engineering and medical equipment. The turnover for the year ending 1994 was £727,200,000. During 1995 (the year before the interview was held) the company invested in various types of advanced manufacturing technology including CNC, robotics, AMH and CAM/CAD. Almost £70 million was spent on capital expenditure during 1995, and about 50 investment projects were approved by head office.

In this company, investing in advanced technology seemed to be necessary in order to maintain world class leadership.

> 'To be a world leader, you do need top class technology and that's the reason why they buy this kind of stuff. But also their competitors are doing it and although they are world leaders, there are two or three world leaders so you do need to have an edge'.

Each unit could approve capital projects up to £100,000. Projects that cost more than £100,000 had to submitted to head office for evaluation and approval. The interviewee was the finance manager at head office managing

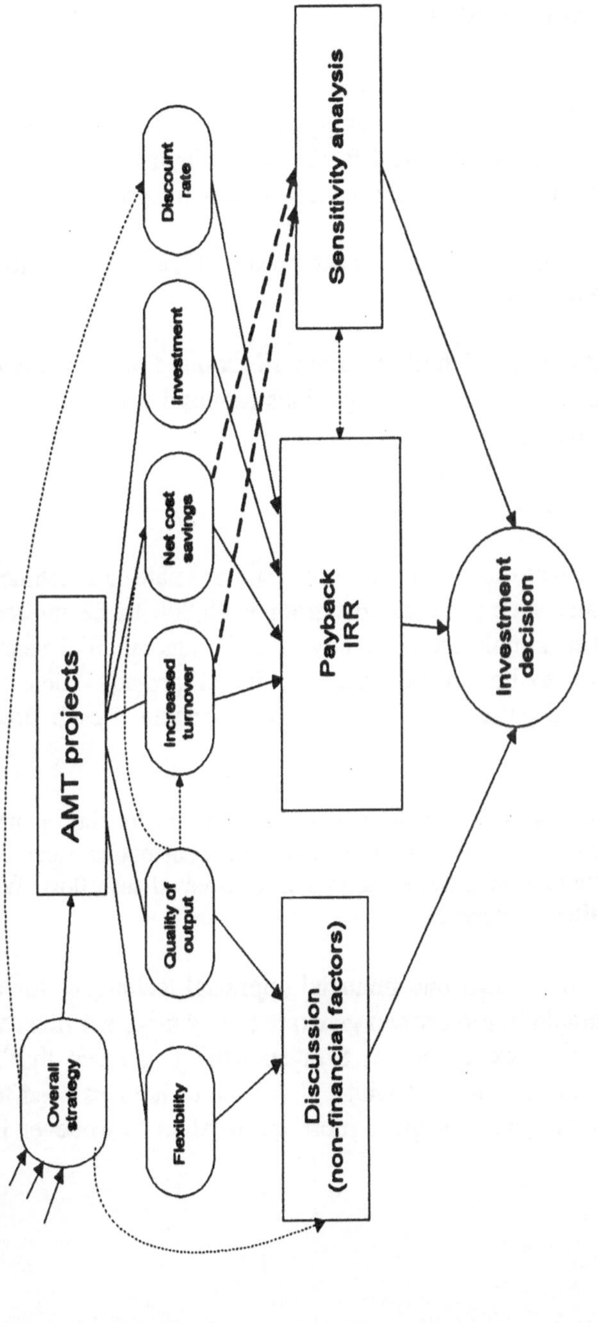

Figure 5.1 The investment decision model in Baby Food Ltd

monthly reporting from subsidiary companies and responsible for the appraisal of capital investment proposals over £100,000.

Developing a Castings Process Project

A recent investment proposed by an automotive engineering subsidiary involved a new factory based on a new castings process. The new factory would be based on an existing development facility where limited numbers of castings were already in production for car manufacturing companies.

> 'The process has just about been, not so much proven, it has been accepted. It was proven a long time ago'.

The new factory needed about £16 million to put all required machines and facilities in place and start to produce castings. This proposal was submitted to head office for evaluation and approval.

Factors Affecting the Decision

Most investment projects that come to head office tend to relate to machines for replacement. Typically such projects cost from £100,000 to £2 million and are evaluated by their simple after-tax payback. In general, five years payback or more is seen to be questionable. For larger projects such as building a new factory and putting in new machines, discounted cash flow methods are used.

> '... Discounted cash flow is our main criterion and we are looking at net present value. We do look at the internal rate of return but not as much as net present value. We need sensitivity analysis on discounted cash flow. We also look at simple after tax payback'.

The company uses more than one financial appraisal technique 'for a balanced view.' For example 'a project can generate a very large net present value [but] if it has [a payback period of] 8 or something like that that's absolutely enormous. It's a long time to wait ...' All the techniques need to be reported in the appraisal paper 'it gives more information to whoever is going to approve it'.

A risk adjusted rate is used in discounting cash flows. It is a single rate in the range 16-20%.

> '... We don't apply a risk to each project, we just have a general risk factor ... It's not calculated every year based on a formula or anything. It is something which has been in place for a few years now, and we look at it every so often for its potential use just like an industry average'.

Also, sensitivity analysis is used: 'to give a measure of where the risk factors are.'

The main risk in the castings process was that 'there was no visibility of particular orders [customers].' To reduce this risk: 'we were looking for order cover for the factory and we had to get letters of intent from various customers before we could go ahead.' There was order cover for quite a substantial proportion of the cost over the next few years. Moreover, to get 'some potential in the market for more orders after that ... we commissioned a survey of what people thought the likely world demand on [for] castings is going to be over the next ten years, and that was quite positive.' The survey results were used as an indicator of a general trend of what the demand is likely to be.

> '... It [the survey] also identifies opportunities in the sense that you look at who you might want to sell to, such as BMW or Nissan or Peugeot. The aim of the survey may be to see what sort of cars they are going to be produced in five or ten years time and you want to be speaking to them now because those are the kind of cars which will come out in five or ten years time which are being designed now. They are thinking about what sort of castings they are going to want and you want to get your foot in the door quite early'.

Another example of the way in which risk had been reduced in an investment project involved a subsidiary company which wanted to produce new products. It was decided to go ahead with a limited programme only, evaluating how successful the first product was. 'Although there is a delay in producing the second one, but it is a sort of reduced risk.'

If problems had been experienced with a particular type of machine, this would receive particular attention if further investment was planned. For example:

> '... A machine went wrong in one of our companies, and another one wanted us to buy another machine - the same machine but a different company. A

> task force was formed of a couple of engineers and a couple of accountants from the relevant businesses to actually look at the new machine, and investigate it in a bit more detail. The company which wanted to buy the machine will have already investigated it and would have come up with their proposal. But in light of what it didn't know previously about the other machine, those aspects were looked into it a little bit further before the other proposal was made'.

Investing in new technology may imply an additional risk, such as the unexpected delay if the technology does not work or the problem of not getting the right technology at the right time.

Further, liquidity may limit the number of investment proposals put forward because head office sets financial targets for every year. The unit should verify that it can afford the required investment from its own resources and satisfy its cash target before submitting any proposal to head office. If a unit requires an investment which it cannot afford from its own resources, good reasons would need to be provided.

It might be enough for a relatively small project to satisfy the financial criteria of payback and positive net present value. However, for larger projects such as building a new factory, fit with overall company strategy is very important.

> '... We set a strategy for each business and that's like a three year period of work. It's important that that investment fits in with that strategy. So that one could come along with an investment proposal which takes the company to a new product and we may not want to go with that project. So while it might make financial sense perhaps, based on the numbers it has non-financial criteria such as why do we want to be in this business, to consider'.

> '... It's a very big investment for the group so obviously you have limited resources and unlimited proposals I suppose, and you have got to decide whether that's the place where you want to channel your investment. So you had a strategy review just before then where what was coming out of the review was that the castings business is an area of growth in the future. So there were good reasons why we should be investing such a lot of money there'.

It should be noted that most investment proposals that come to the head office have a positive NPV. However, in addition to reviewing the fit of the investment with overall strategy, significant consideration is given to non-financial issues such as customer expectations, competition, quality of products and employee working conditions. Some examples of such considerations:

1. Super alloy scrap retrieval - impressing the customer:

> '... We have a company which makes super alloys and there is some scrap which results from the process of making super alloys. I think you could clean this scrap up and sell it on or do something for the customer with it. They wanted to invest in a new machine for about £100,000 or so which would do this job more efficiently. It wouldn't necessarily save any money but it would look good to the customer. It would look like they had a world class facility ... So in strict financial terms it was touch and go about whether we should invest in it or not. But it [the company] was so successful that it was very difficult to say to them ... we are not going to allow you this investment, because it wouldn't send out the right signals ... It has a simple payback of about eight years, which I guess is a bit difficult to justify normally'.

2. Finite element analysis - a marketing necessity:

> '... We spent some money on machines on the finite element analysis, and one could argue that it doesn't give you a direct benefit because you are spending some money on machines which don't win orders. I don't see an order coming ... because we have got such package for the company. However, customers expect to see it, so you have got to take that into consideration ... You talk about the implications, you are also bringing in odd financial criteria at the same time ... They also justify part of that expenditure purely financially they said that we have to send this out for a bureau to do it and that costs £X'000s. If we do it in-house then it costs £Y'000s. One could argue then, do you have to do it at all? Yes we do because the customer wants to see it'.

3. A small business purchase - strategically desirable:

> '... I can think of a small acquisition we have made, that doesn't necessarily satisfy strong financial criteria and yes it is a bit of a payback, but it is marginal ... strategically it was best that we bought this company than not. It fitted nicely and there were some benefits and also in terms of marketing ... It is very difficult to put numbers on to that but the non-financial reasons were quite compelling ...'.

Investments designed to improve quality have both financial and non-financial implications. For example a unit producing cars has invested in a project to improve the quality of the paintwork:

> '... It looks better to the customer and it saves money having to repaint, because [this type of car] has 13 layers of paint ... If you get your quality right at the beginning then you have much less non-conformance at the end. To the customer as well, he is getting a car which is in top condition'.

Investment to improve employees' working conditions may not require head office approval if it is less than £100,000. For example:

> '... One of our units wanted to put a new office building in, and there were very good reasons for it. It's not one of those things which you can necessarily do financial appraising on. You can see some of the benefits like energy efficiency and reduction in rent from the buildings which you were renting previously. You might not be able to come up with the net present value for the thing but you can justify quite a lot of the spend financially'.

> '... You have partly improved working conditions, improved communications. Because they were having business re-engineered they were moving from a functional approach to a process approach and what they wanted to do was get the people from the various departments to start working as a team together, they wanted everyone to be in the same building and they wanted everyone to be talking together. There were obviously some financial benefits but there were also a lot of non-financial benefits such as people being able to talk to one another'.

In this example, the problem was that the unit was not achieving the return on sales that had been set for it. ' So, we said that you can't have this building until you commit to achieving a particular return on sales ... The building actually started off on a kind of a wish-list'.

Decision Making

In making an investment decision to develop the castings process, there are a number of issues that are considered, such as positive NPV, strategic fit and reducing the possible risk in potential demand of castings. In general, more weight is given to NPV. Before an appraisal report is sent to the finance committee/investment approval board, the executives of the unit come and,

> '... we all sit round a table and discuss it and hammer it out, because at the end of the day, no matter how many numbers you put down on paper, it is not exact science, so it's management view or the strengths of forecasts'.

The stability of the business in which the investment will be made may affect the final decision:

> '... You would have to look at the resources of the business then, because if they are not doing well then they are going to be stretched, and you would have to question whether it would be the best use of the resources at such a time. If you are going to invest a lot of money in a business, it's perhaps easier to invest in a business which is stable, rather than one which is going through a lot of crisis and a lot of change'.

The decision is basically based on judgements and trading all the criteria mentioned above against each other. There is no systematic method used in weighting these criteria.

Formal Process and Presentation

The company has 'standing orders', company rules that give guidance on the preparation of investment proposals. Every project over £100,000 should be submitted to head office in a standard format. It is reviewed by the finance manager, and a review paper summarises the NPV, the simple after tax cash flow, payback, sensitivities, project implications, risks, and a recommendation. Only the financial issues are quantified. No attempt is made to quantify or 'score' other considerations which are reported in a narrative manner. In addition to the actual body of the proposal, a form is completed that specifies the project time-scale and summarises key dates.

The proposal that comes to the head office has to be signed off by two members of the finance committee before taking a decision to implement.

The investment decision in this company is summarised in figure 5.2.

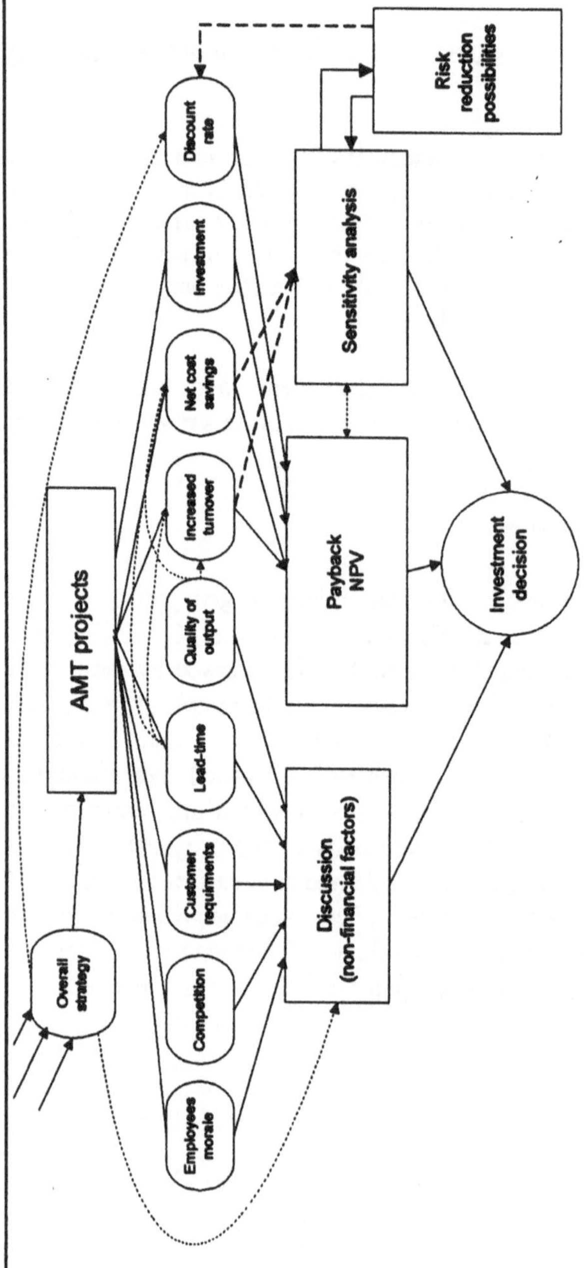

Figure 5.2 The investment decision model in Engineering Plc

Case 3 - Newspapers Plc

Background

Newspapers Plc is a leading company in publishing and printing regional newspapers. It produces a range of titles including three daily regional newspapers, 12-14 weekly paid-for titles and about 75 free titles. Turnover for the year ended in June 1994 was £84,831,000 and 1,617 employees were employed.

Newspaper businesses have a very high labour content, and until recently, working methods have been dominated by union influences. This made it very difficult to introduce new technology. However, the collapse of union power and advances in computer technology over recent years have led to many changes. The company is now a major user of computer technology and both editorial and advertising staff use computers extensively.

> '... Over recent years we have placed increasing emphasis on the use of computers in production with editorial and advertising functions being separately computerised. Now instead of keeping these areas separate and later merging their outputs together by a manual pasting up process, we are doing it all within the same computer system, but it is only recently that this has become possible because of increasing computer power, which means that we have the capability to process colour photographs at a sensible rate'.

Printing Direct to Plate Project

A few years ago there were two fundamental processes in preparing to print a newspaper. In the first process, a Linotype operator performed a typesetting operation whereby individual characters of the alphabet were assembled in the form of a master plate. Then, as a second process, a printing plate was cast in what was known as the 'hot metal' process ready for use on the printing press. These processes were replaced by a largely photographic process. Originals of individual articles or advertisements are still made up by hand and assembled into a master for the overall page or collections of pages. These are then photographed to create a transparency, used in turn to create a printing plate.

The next stage in computerisation was to retain all the individual stories and advertisements within the same computer and electronically arrange them into the format of the individual newspaper pages. This was quicker and cheaper as it eliminated the pasting up process whilst also allowing last minute adjustments to the page layout. This 'full page output' is advanced for the newspaper industry, and Newspaper plc have recently achieved it in several major centres.

A recent investment decision was taken to move to the most advanced stage in printing, that is, printing direct to the plate. By investing in this type of advanced technology, Newspaper plc became the leading company in the UK and is very nearly first in Europe. Within this process, after the newspaper pages have been fully assembled within the computer, instead of a master being printed so that a transparency can be created, the data is output directly on the face of a printing plate. This eliminates the photographic stage completely which produces a saving in materials and time but without degradation of quality.

Factors Affecting the Decision

The capital cost of the 'printing direct to plate project' is about £1.4 million. Its main advantages are: reduction of processing stages, elimination of human intervention, and obtaining high quality with reduced chances of error. Two main factors dominated the decision to invest in the new technology: cost reduction resulting from savings in labour and material costs; and improvement of product quality.

> '... We save some manpower and also printing material but we substitute that with increased depreciation. Simplistically, these two factors almost offset each other, however, depreciation is a non-cash cost which also will not inflate with time and therefore there is a benefit, and in addition, we are improving the product ... That is a satisfactory basis for investment'.

New technologies have had major impacts in printing and media:

> '... The newspapers are the established market leaders and therefore new competitors, Auto Trader being a classic case, would attack particular sectors of the established market, where margins were thought attractive. They were able to do this by the use of new technology which enabled new products to be introduced much more cheaply than previously. To compete, we had to improve the service given to advertisers and also to reduce [cost], a key factor in this is the elimination of high labour content in traditional newspapers'.

For any investment project to be approved, it should show a short-term return. Although a strategic project may be felt to be right for the long term and the board would be 'sympathetic' to it on a grand concept basis, its chance of being approved would be much enhanced by a tangible short term benefit.

> '... Within the business we previously had a culture which was reluctant to spend money on capital projects because of the perceived risks of making expenditure, however, we changed this by encouraging general management to bring forward items which had a quick payback which we could feel comfortable in quickly approving. This helped management to think constructively about how they could improve their areas of operations'.

Two methods of assessing return are basically used for financial evaluation: payback and ROI. ROI receives the greatest emphasis in decision making:

> '... The return on capital is what [is] happening to cash flow in the bank and that's everything. So I tend to think in terms of investment and tend to stop focusing about less than 20% unless there was a very persuasive long term strategic argument'.

Payback and ROI are used mainly because they are easy and simple in use. 'Sophisticated calculation [DCF] in investment appraisal is generally not found to be very useful in coming to a decision.'

Prior to considering the financial evaluation of a project, the project should be discussed in the context of its role in the business and the overall business objectives. If the project is thought to be consistent with the long term aim of the business, then its financial return will be considered.

Sometimes, capital expenditure requests can be initiated as a result of direct customer requirements. For example:

> 'A particular important customer could wish for various improvements in the printing of its magazine and to retain the order we would need to look towards incurring capital expenditure. In the first instance, of course, we would wish to protect our present business by making the expenditure. However, in negotiating with him we would seek to obtain assurance that enough work is guaranteed by the investment to more or less cover us over a reasonable time period. Ideally, there should then be the opportunity of using the enhanced production capability to improve our sales to the market place in general, that is, by providing us with a secure foundation of work, the primary customer gives security to an investment which is then used, in part, to expand sales elsewhere'.

The only project which would be likely to gain approval 'without showing a proper payback is health and safety.' However, these projects are small items with a typical value of about £5000.

The basic risk of being the leader when investing in new technology, for example the direct to plate printing project, is that 'you might be caught on the wrong foot by picking the wrong avenue of the technology, that's normally a big risk.' The risk has to be covered by the manufacturer to the extent that a guarantee is given that new technology will actually do what Newspaper Plc require.

> '... This printing direct plate installation was the first in the UK and just about first in the World. We therefore insisted that the supplier set up a pilot operation for trial purposes on our site. This is now running on an extended test basis so that we have the best part of a year of trials before we have to make our final decision. We will only pick the new technology if we are sure that it will work and have obtained from the supplier an agreement to provide us with the old technology in its place should it prove unsatisfactory. So, we have covered ourselves as much as we can'.

> '... It's risk to the manufacturer ... The risk if it doesn't perform is back to him. The danger of course is that we have wasted time exploring the wrong route and might find it difficult to change technology to a more secure basis because of the time constraints of the overall project within which direct to plate is only an element'.

Decision Making

In deciding on the particular model of press for the future, the operations manager and his team visited manufacturers throughout Europe.

> '... They came back with details of several manufacturers' products and had an analysis which had weighting's and numbers scores ... This said basically, be logical in your decision. That's what they did. They listed all the key criteria and we actually, as directors, gave proper weighting to that report because they had gone systematically through the various points as to which press we want ... These reviews of individual press characteristics were not relevant to the decision as to whether we wanted a new press or not, but were invaluable in assessing which particular press [supplier] best suited our needs ... For fairly common items like computers, we will have preferred suppliers whose technology is both compatible with previous purchases and also fits in our discount structures'.

Formal Process and Presentation

The planning begins with overall thoughts of: 'where is the industry going? What's the best direction our company can take if it is to succeed within the industry?' These long-term views form the basis of an overall business strategy within which individual year business plans are drafted and agreed.

Individual units prepare their basic business plans, and these are considered initially by the financial and commercial directors for consistency with long term strategic objectives. Provided that the individual units are aiming to deliver what is required in overall terms, then they will be given considerable freedom in their detailed implementation plans.

In the case of major capital projects, the case is generally initiated by a senior manager or director with particular interest in that area. For example, the printing direct to plate project was raised by the operations director who produced some rough costs 'which showed that there was some sort of financial sense in it.' 'What he is doing is mc formal signalling to the Board that he has got an interest in that.' e idea gained a broad acceptance and a study group involving both specialist and financial staff then researched the project and came up with an estimation of capital cost and a detailed evaluation of the benefits.

After capital expenditure is sanctioned, regular reports on the progress of implementation are required by the main board. Actual expenditure against sanction and the achievement or otherwise of the required return is reported in a format that is tailored to the particular case.

The investment decision in this company is summarised in figure 5.3.

Case 4 - Tobacco Ltd

Background

Tobacco Ltd is a British company that produces a wide range of tobacco products, distributed world-wide. Its turnover for the year ended 1994 was £5,338 million and it employed about 36,000 people. The company has invested about half a billion dollars in highly automated technology. There are three main elements of advanced technology: highly automated robot manufacturing equipment that produces large quantities of cigarettes per minute, automated material handling that delivers 'wrapping material', the material which goes round a cigarette, and 'filter rods', the filter at the end of the cigarette; then an

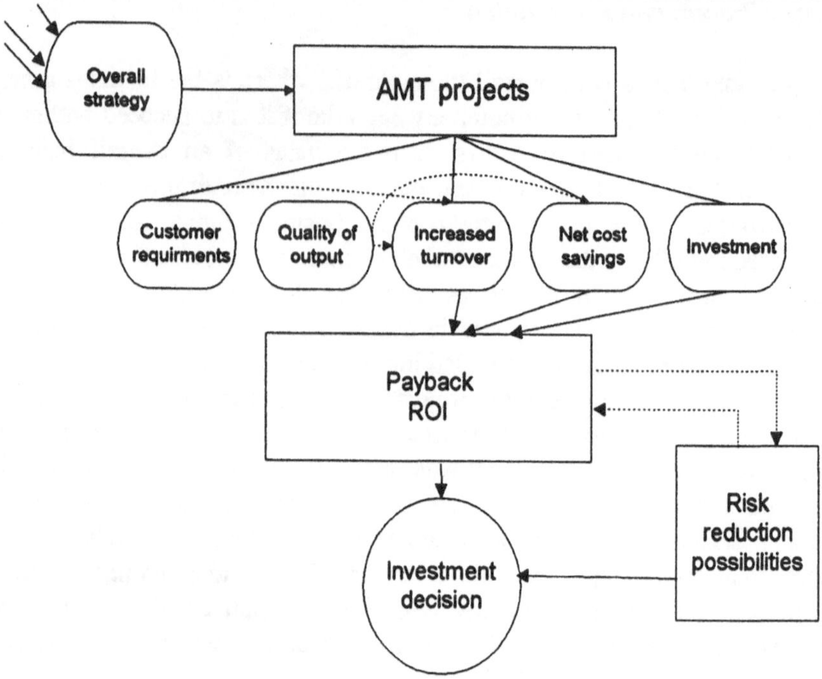

Figure 5.3 The investment decision model in Newspaper Plc

automated vehicle takes the finished products off the machines and puts them in containers; and a 'factory controlled integration system' links all manufacturing and handling units together.

In the late 1980s and early 1990s, a lot of protected and inaccessible markets were opened to western products. Markets such as Eastern Europe, Thailand and China began to grow. So, the tobacco industry 'moved from a kind of industry which was in marginal decline to an industry that's now growing again.' The British site of this company produces only one brand of cigarette which is not sold in the UK. However, even though it is sold around the world, it is very important that it is made in England, the 'made in England' label is very important to the value of this brand. The growth in the brand has given the company an opportunity to expand its factory site.

The company has invested largely in advanced technology because it produces a high quality product in high volumes. 'If you are going to get class "A" manufacture, then you have got to have integrated systems ... effective, efficient flows etc.'.

Factory Expansion Project

As a reaction to changes in the tobacco market, the company has considered an investment decision for a major expansion of the existing factory. The cost is about £200 million of which 25% is building work and the rest is plant and equipment for the manufacture of cigarettes. A vast proportion of this is in advanced technology. It was decided to grade the project into four main phases; phase one was the whole of the infrastructure and the first part of the manufacturing unit and phases two, three and four were basically additions to the manufacturing capacity. These phases were to some extent repetitive. 'We have just finished phase two and we probably won't start phase three for another two or three years. So, it's those phases which allow us to react to market'.

Factors Affecting the Decision

In principle, two options were considered: either to expand the existing factory or to go off-site. 'It was a strategic decision that it was better to expand this site.' The reason was mainly because 'most of our skill is in [this site], we have been there 80 or 90 years. It's very important to us to be able to retain those skills.' Also, when the project was implemented the environment for new development in the manufacturing sector was not good locally. So, it would be quicker to upgrade this factory than to go out of the county and seek land to build. One of the limiting factors was space because the factory is located close to residential areas. Limited space led to a certain (compact) type of technology being chosen.

The maximum output that could be delivered on the site was an important consideration in the decision to expand the factory.

> '... Checking the output, which was important, because if we perceived over the long term the requirements of the market are going to exceed that or come close to it, then we don't want to be in a situation where we can't meet that forecast. Before we decided whether or not to move out to build a new factory, we spent a lot of time discussing the maximum capacity. Quite a lot of discussion was on if we take this decision today, what does that do to our long term capacity? It might be a cheaper option over [a] five year forecast,

but what about in 12 years time or in 10 years time? Will it constrain us? … In our business, where you can't satisfy customer demand is the absolute worst thing you could possibly do. So it is very important to us to understand what our maximum output could be and where the market is going'.

Before going into financial analysis, a number of strategic issues were considered, such as 'how the factory should be developed, what choices are involved, what type of machinery do you think you would want etc.?'

A major investment, such as expansion of the existing factory, should begin with 'a view of where the market will be in 10 years time. Although it's a growth market, it's quite volatile.' The market is very competitive, so the ability to compete determines company share. Initially, analysis was in terms of the group as a whole. There were opportunities in various markets and estimates were made of potential sales, and of what the sensitivity and volatility would be.

'... We then did a financial project to meet that volume, to give us the ability to turn it up or turn it down and to put in place a number of decision points where you could decide to continue to invest ... to delay the investment or ... to cancel the investment'.

The project was analysed financially using discounted cash flow and its impact on working capital, profits, how long it took to get the cash back, and return on assets. 'But in essence, the net present value was the thing which really drove it and obviously the up-front investment [was the] liquidity in that.' Liquidity was looked at through payback using discounted cash flow (discounted payback period). Net present value is the preferred financial appraisal method because 'that is the way you add value, that's what company values … Cash is what drives the value of business'.

It should be noted that the financial analysis was not 'the thing that came first'. Many other issues are considered in the decision:

'The financial analysis is one element of the decision, it is a very big and important element, but I don't stand up and say - this is a project which has an NPV of xxxx - right we should invest. I stand up and say these are the benefits, the costs, the risks, by the way it has this return, this is the cash impact, these are the reasons why you should invest. In this case you should take a lower return because it has these other benefits'.

In a growing market such as the tobacco market, flexibility becomes an important issue in terms of the volume that the company can readily cope with. Flexibility also means the ability to change product lines.

> '... We have actually got a proposal here which has reduced our ability on the second line. You have the high speed, high volume, high quality output, but it's not as amenable to changing products throughout the factory ...'

Quality of outputs is an over-riding issue in any decision in this company. This was stressed as follows:

> '... We need a high quality product to fight off competition ... more important than anything else is quality. We would never do anything to impact on the quality of our output... This is a consumer product. It's a high priced product ... It goes to the far east. It's a very high valued product, quality is absolutely paramount ...'.

> '... When we look at our targets, our targets are high quality manufacture and the next one is low cost manufacture and not the other way around. So we don't say the first imperative is to get your cost down and the second imperative is quality. The first imperative is quality. Certainly people don't come in and say, this is a quality improvement therefore I want to spend this money, it has to be justified. We have to see a cost-benefit, you don't authorise "spend" as an act of faith. If someone can't explain or communicate the benefits of the cost then they don't get it. If they don't have the money, then we don't invest'.

> '... We are investing in high speed technology, we are investing in computerised delivery equipment, control systems which should enable us to reduce our cost and make the process more efficient. At the same time we are able to deliver to the consumer the product which they want. So essentially we are buying the ability to manufacture high volume of good quality product at low cost'.

When investing in this new technology, two main sorts of risk were involved: 'what happens if the technology doesn't come up to standard and what happens if the supplier fails.' To reduce the possibility of technology failure, efforts were made to get evidence from elsewhere in the world,

> '... where someone has used it either in our industry or other industries or in our competitors. So, we did visits when we chose the automated guided

> vehicle system. We visited factories around the world to find out how it worked. For example, in automated guided vehicles you could get wire line technology on the floor, you can get ... radio controlled from beacons which are hung from the ceiling. We went for wire line because it was proven and we felt comfortable with that'.

Also, some simulation exercises were undertaken on material product flows around the factory before deciding on the layout and how many vehicles were needed. For the risk of supplier failure, a particular type of contract was done.

> '... because it is a new technology so it hasn't got standard terms ... We try to look at how we could get risk to that party most capable of carrying the risk'.

All risk issues are discussed as a team. They are reflected either through 'how you allocate the risk' or through sensitivity analysis and adjustment of the discount rate.

> '... In essence you may use a number of discount rates. What are you trying to do? We try to move away from "I must achieve a 15% return" because all that happens is that people say well what do I need to do to the model to make it achieve a 15% return and people push things around. They change their assumptions and their inputs so that they get 15% because they can see that as the hurdle. What we tried to do was to give our board a feel, an understanding, of the scale of investment, the likely return and the sensitivities of the returns, should things happen in the market or should things happen on the project and an understanding of the overall picture so they can make a decision. For example we could have done a chart like [see below] and said this is time and you have got your volume scenario ... They could feel confident that we are going to be above this line, then they would want to invest. Then there would be all sorts of sensitivities that we would have done i.e. what happens if the price moves; what happens if we don't get the volume and there is a delay in the project; capital cost works up, there are all sorts of sensitivities, so that they could feel that they had a high degree of confidence. Even if we had these things that went wrong, it's still an attractive investment. We did NPVs as well at one level but the way we presented it, the risks in here were the main point of this investment where the volume falls away, so what's the risk of losing volume'.

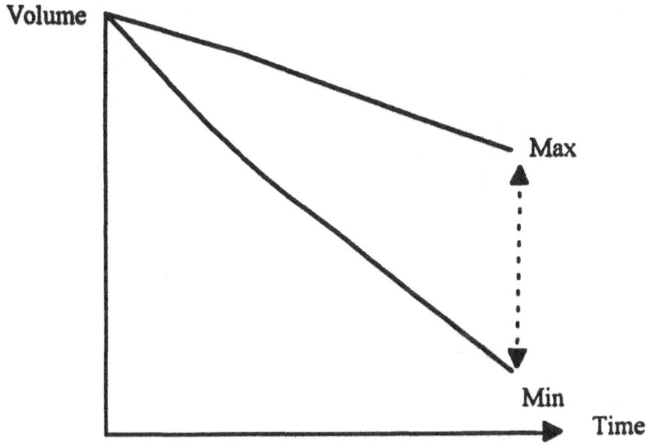

Decision Making

Determining the level of technology that the company should invest in was an important decision because 'as you invest more and more in advanced technology you actually get a reduced return [see the following chart]. The question is where should you be on the curve.' Given that the company had to work within certain specifications regarding utilisation efficiency, product quality targets and limited space, working with suppliers was an important element in deciding the technology level.

> '... We had to make decisions as to whether we wanted a machine from that supplier or that supplier. If we get it from that supplier then we need to have this type of sort of automated handling equipment ... We looked at three different machinery layouts because it depends on which supplier you go to as to what you need. We did comparisons in terms of financials and also in terms of what the engineers felt they could deliver, what they could perform'.

Formal Process and Presentation

In January 1991, a team was put together to consider how this factory could be expanded given the potential growth in the market and whether it would be feasible to do it on the existing site. Six months later the team reported to the board and the project was approved.

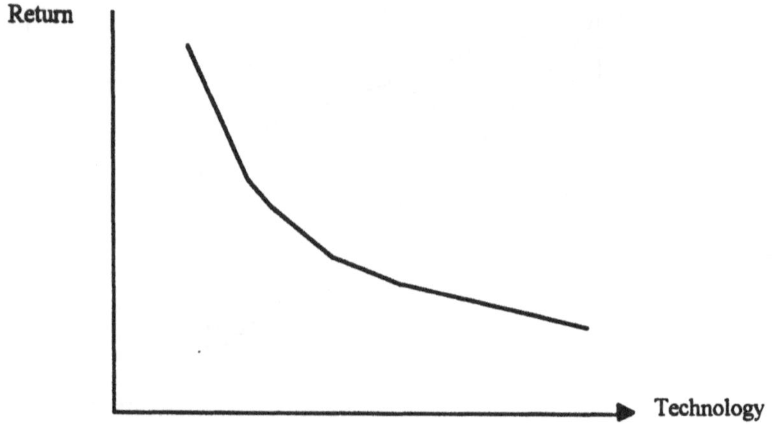

> 'Periodically, over the next four years we went back every year with a review and recalculation of the investment and everything else. To some extent within that total investment, you then had local authority and the board authority'.

Whilst different projects have different levels of authorisation (depending on their value) the evaluation process is similar for all projects:

> 'Whether for a £200 million project or a £10,000 project, you have to report about what are the cost/benefits of that investment, and what are the risks of that investment. That means you do all sorts of models and sensitivities and everything else'.

Also, a sanction form has to be completed for certain types of projects.

> '... For certain types of projects we would expect to do cash flow analysis but [for some projects] there is more about what ... the project is going to deliver. So for example, if we have got a Health and Safety investment, then in that instance you are thinking is that the best way of making that improvement? Do we need to make that improvement? Is that the best way of achieving it? I wouldn't go into net present value analysis necessarily in that, where you haven't got an income stream ... We would do all sorts of analysis on how best we could deliver that benefit'.

The non-financial elements in the decision 'are discussed as a team'. A scoring method is used only in environmental decisions. In such decisions:

'... we look at various factors such as cost which is not the biggest issue there, we are more concerned with the level of recycling we could get, or our good neighbourliness. We have just built a bio-filter which is essentially a large peat bog ... which a lot of our odour is pushed through. That's an investment which is purely driven to ensure that our neighbours see us as a good corporate citizen'.

The investment decision in this company is summarised in figure 5.4.

Case 5 - Packaging Plc

Background

Packaging Plc is a holding company consisting of 140 world wide companies with its head office in London. The turnover of the group was £2.45 billion in 1995. The group has a variety of activities such as coating, graphics, metallised paper and film, and medical and pharmaceutical packaging.

Investing in New Technology

The company has grown very rapidly by acquisition (75% of the company now is the result of acquisitions) and, while the holding company has grown rapidly, many of the companies within the group have never worked together before, even those which have been in the same area, and some may even have been competitors.

The company has invested in different types of advanced technology. For example, industrial saws are now sold with intelligent video screens/control boxes. If there are problems, customers can dial in directly and a video shows what the problem is. A video screen is also used to show how the problem can be solved, and accordingly, service costs are much reduced.

In the packaging business, advanced technologies have improved the quality of many products, for example, in packaging for ice creams, butter, margarine and yellow fats. Although the packages are thrown away, the level of technology involved in producing thin walled cartons is very sophisticated: they must not only be light and cheap, but also attractive to potential customers. Absolute uniformity of quality in production is very important because, if a customer's production line is stopped, the problem

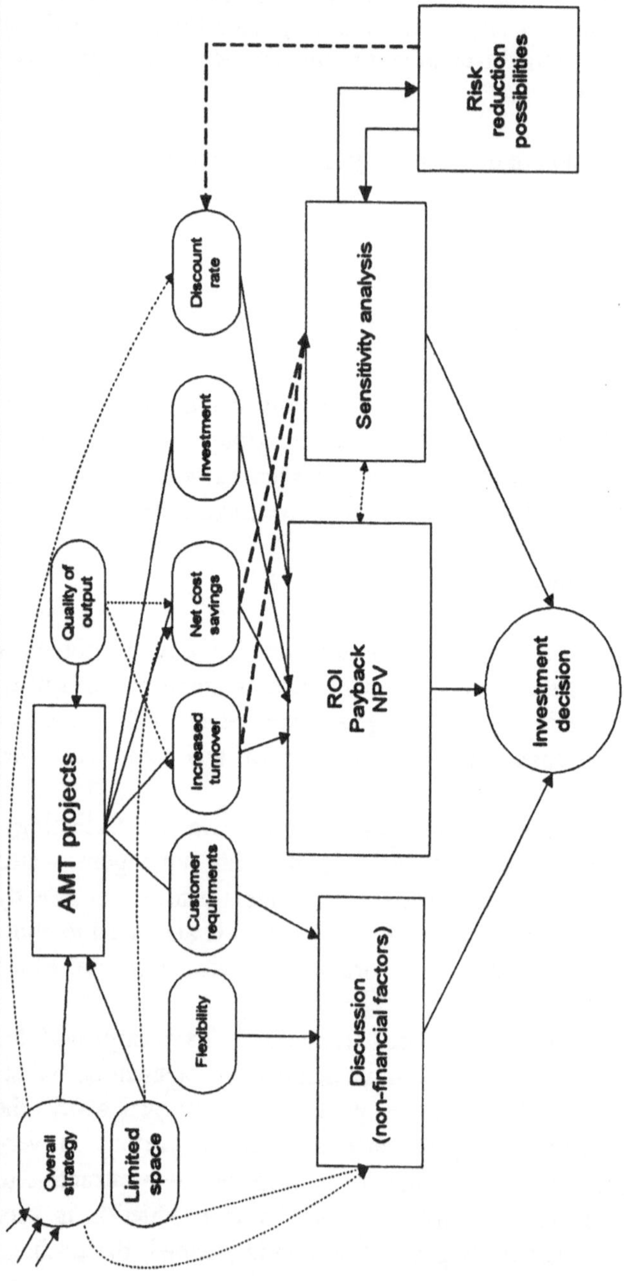

Figure 5.4 The investment decision model in Tobacco Ltd

could be very expensive. Thus, the output has to be of the highest quality. An investment was made in an automatic inspection system, which automatically re-programmed the machine if there was fault. This system did two things: it aimed to guarantee customers 100% for quality of their printing and decoration, and it increased employee productivity.

Advanced technology is also important in the packaging of fresh or chilled goods with long shelf lives. These are packed in an inert atmosphere, then at the point of sale, oxygen is introduced to turn the purple meat to red. A lifetime of around 21 days can be achieved by using such modified atmosphere packaging.

In 1995 the company spent about £24½ million on R & D across the group. Traditionally the philosophy of this company has been that each individual site is independent with its own managing director, board of directors and profit and loss account. However, the company is about to change this philosophy by gathering clusters of companies together into divisions which report directly to an executive director[4]. There is still a strong philosophy within the company that 'centre is bad and local is good and you have to row your own canoe.' Thus, the corporate technology budget is not very large.

The only sizeable amount of corporate money (about £450,000 per year) is being spent in the area of Knowledge Based System (KBS). This started by trying to identify rules about how people did their designs. The systems are open, continually evolving series of 'what if' questions or answers to 'what if' questions. The same kind of software can directly integrate the design system into the appropriate manufacturing process.

> 'So you could say down which of our lines would this object go, and how much would it cost if it went down these different lines. How fast would it go down these different lines. How much would it cost if we made it out of polypropylene instead of polyethylene. How much would it cost if we had a width of 1.8 mm instead of 2.0 mm and so on'.

The KBS allows information to be captured and stored.

> 'The knowledge of the company is encapsulated and uniformly applied to all objects that you are making. It does mean that you have a body of knowledge which is people independent so if you have got somebody who is 63 years and who has designed all of your lipstick cases for the past 30 years and he is going to retire, then you can get his brain and his rules onto the machine'.

The Investment Decision in Context

The corporate level of Packaging Plc encouraged operating companies to invest in KBS by providing them with an opportunity to pilot the system. 'We can then offer them advice as to what software to buy; how much it will cost; capital proposals to help them create one.' Currently, seven operations have spent a total of about £450,000 themselves.

The KBS can help to take advantage of the increasingly transparent nature of the supply chain:

> 'We have said that we will integrate some things within the company ... So, we can put in all the rules concerned with our customer filling lines. We could then design an object together which will satisfy the customer filling lines rules and our production rules. So it is immediately transferable ... It's an enabling software which in principle does enable us to seize more of the knowledge cake at either end and manage ourselves and manage our suppliers and our customers better. In this area we are ahead of the game, certainly in the packaging'.

> 'We are doing some tooling for our own proprietary technology which is called rotary thermoforming. There is a business where we are making profiled aluminium or cardboard. A company [customer] asked us for a design, we sat them down in front of the computer and said "... oh you want a name like that?" "Yes, something like that, maybe change it like that. Now if we ask you to produce that are you going to take a couple of weeks to do the tool for that?" "Oh no that's already done!" Essentially, you can pass the information down to a tool designer from there and get the tool made up. If asked about the costs of it, then we can show the costs already there on call. We walked out with an order - just like that!'.

The corporate level must believe in convincing the operating companies to invest in advanced technology. The introduction of KBSs involves personnel issues between and within both corporate and plant level operations. Culture change must be accepted and planned and the investment must be justified financially:

> '... He [the managing director] has to say, I'm prepared to buy some hardware which might cost £25,000. I am prepared to train two or three of my guys which is going to cost me another £10,000-15,000 . I am going to buy 5 sets of this software which might cost about another £20,000 and, and, and ... and in doing this I will cut my time to market from the existing 2

> months down to 2 weeks. I will be able to carry out 50% more jobs than I do at present or I will have the opportunity to lose some people. I will increase my security and control that I have over my business because I will be incorporating all the knowledge which is in the business into the software. Those are typical statements which will be made ... In fact, we did actually go out at the end of last year to the piloting companies and check how they were doing so far against where they said they would be in their capital expenditure requests. I think all of them were where they said they would be or just slightly ahead of what they put in their capital expenditure request'.

Over the last 3 years the company has invested between £130-190 million a year, but this year 'we gave two profit warnings and it became very obvious that we were also making a statement that we were cutting down in capital investment at least this year.'

Return on sales was introduced by the current management in 1987 as the key financial parameter for investments at this company. Prior to that the key had been return on capital, return on operating assets. Current management have tried to increase the volume and quality by appropriate investment and setting the return on sales as the key parameter. People have been rewarded on return on sales as opposed to return on net operating assets. Accordingly, the return on sales was used as a key financial parameter for approving capital investments. However, there is a move to change the performance evaluation method from return on sales back to return on net assets[5].

Investment projects are not normally judged on the basis of discounted cash flow methods. DCF was used ten years ago 'but it seems to me to have fallen out of favour a little bit.' If NPV is calculated, a discount rate of 8% is normally used.

The payback method is very important in evaluating investment projects. It may be the only reason to reject a project:

> 'There was a very interesting project where ... a paper project, in which we could really take over a total packaging solution for one of our customers ... Some of the units would do better than others out of it, so there was a certain question about transfer costing and all the rest of it. But the thing which really excited [worried] everyone was when you looked at the amount of capital you were supposed to be able to put into it, the payback period was getting out about five years ... Yes, we are very excited but look at that payback. We are really putting a lot of money out on the line ... You are putting a lot at risk for that. So, of course payback is important'.

The commercial and technical risks are identified explicitly for the project according to the company rules. 'If there is a high risk you can pilot... Certainly we do try to test by "doing" not by "talking".' Many information technology projects had gone wrong and so the company now encourages its managing directors to invest by helping them in pilot schemes before they commit money, as in the case of the knowledge based system.

The most important two reasons often used to approve an investment project are customer demands: 'we won't get the business without it', and competitor action: 'he is already a year ahead of us and we absolutely have to do it.' These are the two reasons 'that carry most weight'.

The non-financial issues are reported and discussed in the approval process but only the American units score these issues. 'They say oh this project is worth 78.5 points and this one is worth 83.6.'

The coating business in America created a programme called 'accelerated commercialisation of technology' (ACT), a system to be used as a part of the 'company's self-assessment toolbox'.

> 'What this is doing is providing techniques for gathering together all of the available information for all of the projects which are around, either coming from the exterior or coming from internal sources ... It's a multi-departmental thing which everyone has to get involved in, finance; marketing; production and technology. Really trying to provide an intelligent filter ... The ACT progress does set up a whole series of milestones as staging posts against which a project will be judged against a number of different parameters, financial being one, technical being another and so on. Somebody has to make some judgements explicit about what the project can offer, what sort of return it is going to generate; whether it's going to generate new business; whether it is going to replace old business; what its risks are; what its risks are technically ... an explicit attempt made to tie together business strategy with technical strategy or marketing strategy'.

However, because:

> '... everybody's business is different. So each business then took away the generic bit and made it appropriate to their particular bit of the business. The fact was that they all did it in a completely different way. All separately invested in software consultants to write their own programmes.'

This programme is used in American businesses, but it is not used in Europe.

The amount of the investment determines who can approve it. The executive director can approve investment up to £400,000, the CEO can approve up to £1 million, and above £1 million the investment has to be approved by the main board.

The investment decision in this company is summarised in figure 5.5.

Case 6 - Heavy Engineering Plc

Background

Heavy Engineering Plc is a large company which designs, manufactures, constructs and services capital plant for energy conversion. Its turnover for the year ending December 1994 was about £740 million. The company employs over 14,000 employees.

The primary reasons for investing in AMT are cost savings, efficiency savings and securing a competitive position in the market place.

Two different investment projects were given as examples to demonstrate investment decisions in the company: a long term and a short to medium term project. The long term project involved product development with investments in building, plant and machinery. The second project was a straightforward upgrading of machine tools.

Product Development Project

This £15 million long term project aimed to develop an industrial power machine using aero-engine technology that could be marketed in the small industrial power generation sector. There were three basic choices: manufacturing in the UK, manufacturing outside of the UK, or out-sourcing. The choice made was to manufacture the product in the UK.

A new aero-engine has only been developed in the last four/five years, so this was the first opportunity to apply it in an industrial power application. The decision to apply this new aero-engine technology to industrial power plants was a strategic decision for the company.

> 'There are two key criteria. In terms of territorial investment it is where can we make it most economically in the long run? I guess initially, how can we economise on the investment that we would need to spend money on? In

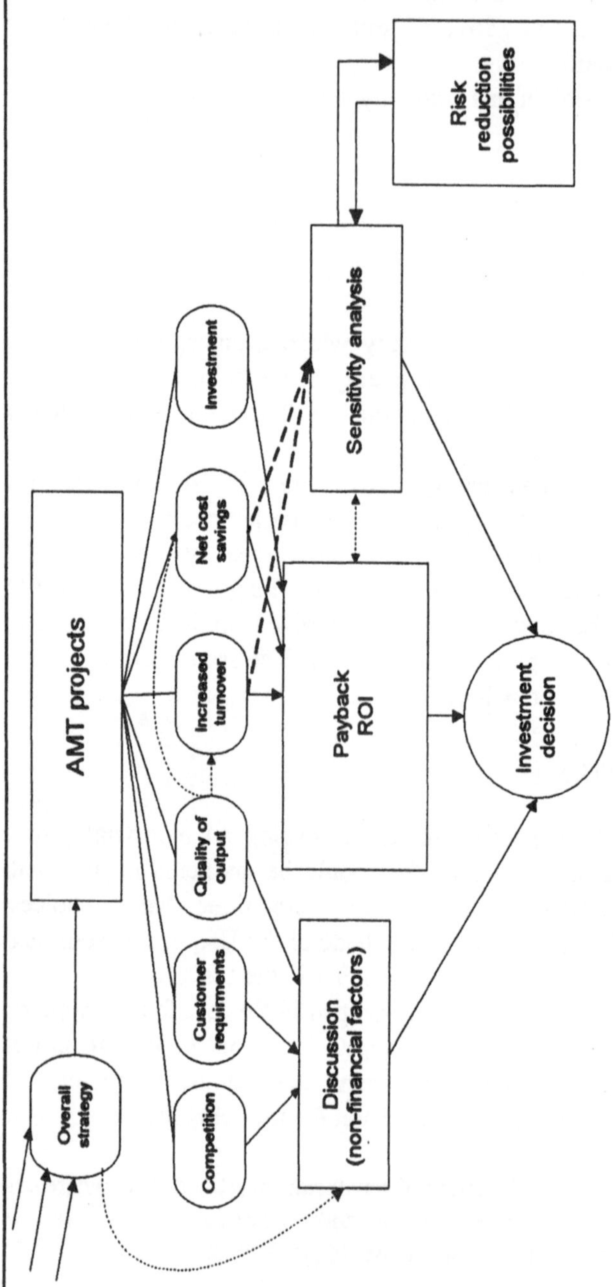

Figure 5.5 The investment decision model in Packaging Plc

other words, can we get Government support and grants? Those are the two key criteria'.

Two main financial appraisal techniques are used for evaluating investment projects: payback and accounting rate of return. Preference for each technique depends on the impact of the investment on the business. 'In terms of the short/medium term we would look at payback particularly if it wasn't on a strategic product related investment ... If it is a strategic thing, then the [accounting] rate of return would take preference.' Payback in three years or less and/or a minimum of 25% accounting rate of return is required.

Marketing the product was a critical issue in the decision. 'Can we sell this product and if so, how many and at what sort of margin?' A risk analysis was undertaken, investigating sensitivity: 'particularly on pricing issues; number of units and on margin costs ... The range of sensitivity, one looks at the downside anyway, is about 10%.' Since there is risk associated with the project, a high return would be expected.

If the risk of a project is too high, the decision to invest depends on:

> 'the overall strategy in terms of the product profile which we have on offer in the particular power sector. Then of course we address the risk in terms of what strategic angle is the company proposing to take in that particular sector. If the strategic decision is that we are going hell for leather, then unless the returns on the project are pretty abysmal, the project will invariably go on ... because issues in the long term like, are we going to be a player in the sector or not, is quite an important issue when it comes down to it'.

Upgrade of Machine Tooling Project

This project cost about £4-5 million aiming to take advantage of continuing improvements in machine tool technology: 'the control mechanism on the machine is improved, computer controlled touch is improved, and over a period of time you get technology which makes the machine go faster and do a lot of things in one operation. Clearly, once you have made an investment, you tend to run the machine for 7 or 8 years before you decide to look at reinvesting.' Basically, the decision was: 'do you reinvest and buy a completely new machine or do you upgrade the existing machinery?' The decision is based mainly on 'what return do we get on each of those?' 'Does it have an impact on, for example, reducing manpower requirements? How quickly can we push product through the operation?' Reduction in material has no effect in this decision.

The effect on the quality of output was important and could be quantified in terms of rejections. Flexibility was also an issue but it could not be quantified.

The sort of technology associated with this machine does not change rapidly. Typically, it takes a minimum of 5 years before the next step in technology. So, technological risk was limited and risk analysis concentrating on demand. 'We would look at the extent of the likely demand, particularly over the next 3 years.' Sensitivity analysis could lead to consideration of the lower risk option. 'Instead of buying a new machine let's look at the upgrade option.'

Decision Making

The decision to invest or not is subject to various factors, some of them quantifiable and others not.

> 'The sort of driving criteria are always issues like financial. Let's quantify as much as we can, especially the key decision point areas. But, at the same time let's look at all the non-quantifiable issues, rank them in terms of priorities and get a feel for the effect of strategic implications. You know, productivity, manpower issues and then try and make a balanced judgement'.

> 'If the arguments are not clear cut, particularly on the subjective issues, then it would begin to beg a question on the credibility of the numbers which are being put together'.

Some issues may play an important role in the decision to invest overseas:

> 'One of the issues which we are facing is, should we manufacture here or should we set-up a joint venture in a low cost country. That's becoming a feature in some of our thinking ... If you are trying to sell a major project overseas then one of the issues is how much local contact have you got in that country because we don't want to finance in their currency, therefore local contact becomes a big issue. One of the ways of increasing that is to set-up a joint venture there. Let's transfer some of our basic knowledge, not the advanced technology, and that way not only helps us with the local

project, but we could use that to help some of our other manufacturing companies'.

The track record of the person who asks for the investment may have an effect on the decision, but this is only:

'... as a sort of final check ... For example how much investment has gone into that unit before; has it actually demonstrated achievement in terms of what it says it can do. Because what we have with a major project is subsequent audit check once the investment has been completed. Has the management got credibility?'

The company does not undertake formal scoring to consider the different issues in the decision.

'The quantifiable aspects of it are obviously the [accounting] rate of return and sensitivity, but everything else is subjective and based on marketing information. The information on a lot of these issues is very sketchy, and it's largely internationally based. A lot of these products are sold overseas, therefore where is the market going; what is the likelihood of the extent of the growth; a lot of subjective really comes into this. Therefore whilst you have quantifiable returns on the one hand, you clearly have subjective information on the other hand. At the end of the day, somebody has to make that sort of commercial judgement'.

Formal Process and Presentation

The choice of departments that usually participate in the evaluation of investment projects depends upon the particular area of investment. For example, for a machine tool replacement, the decision would be manufacturing oriented with input from the relative sales area in addition to the finance department. If it is a project like the new aero-engine: 'a lot of people would get involved, not just the manufacturers, there would also be people like the marketing guys'.

The discussion regarding all the factors, whether quantifiable or non-quantifiable, is reported in the investment evaluation document. The long term projects are approved at the highest level of the company: the board of the public limited company.

The investment decision in this company is summarised in figure 5.6.

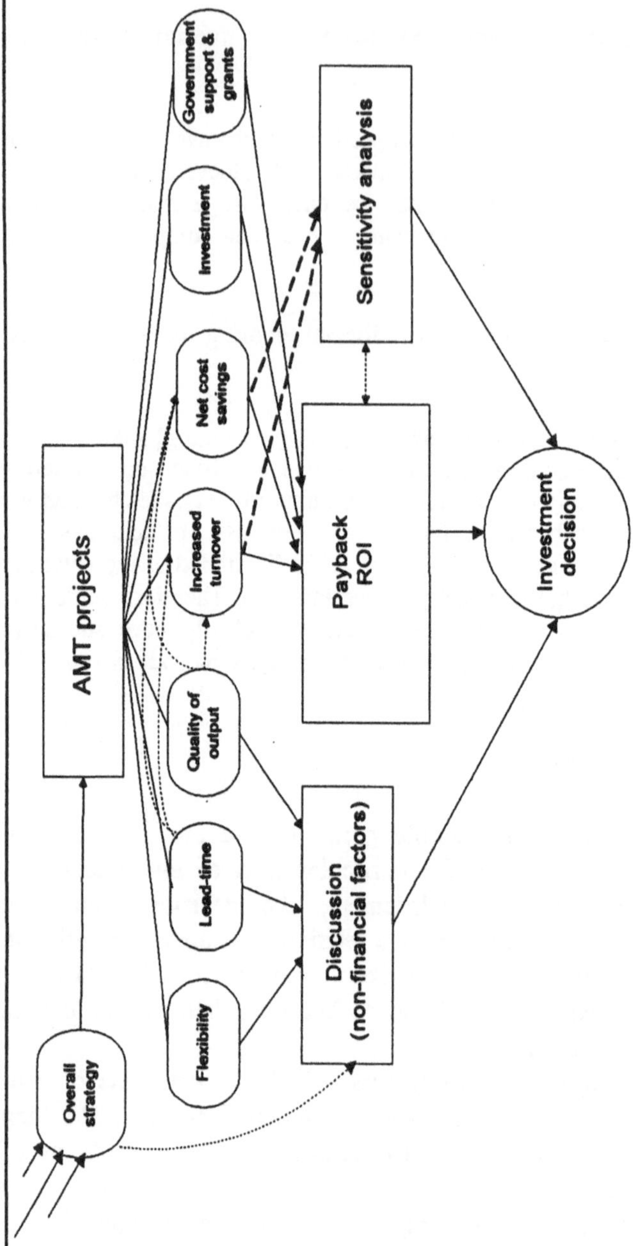

Figure 5.6 The investment decision model in Heavy Engineering Plc

Case 7 - Maritime Ltd

Background

Maritime Ltd. is a leading company in the maritime manufacturing sector. Its main activities include merchant shipbuilding, hydraulic and mechanical engineering and manufacture of marine spares and equipment. The turnover for 1994 was about £253 million, and the company employs over 3500 employees.

Proposed Replacement of an Old CAD System

This is a proposed project to replace an old CAD system with a new distributed processing CD system. Currently an IBM mainframe computer system, CADAM, is employed[6]. The proposal is only tentative and the project is being evaluated in order to prepare a final proposal which will be presented to the board. 'We are currently in the business at the moment of looking to see where we should be moving, and what justifications and parameters there are for making this decision.'

The reason for making this investment is in response to market needs.

> '... We see ourselves, as entering into a phase where the shipbuilding market will be doing one-off special ships. It would be a requirement for us to turn round the design of these issues of engineering and so on. When you are doing a very small series of one-off ships then the question of re-work and so on always looms large in your mind, together with all the lead times etc. So what we effectively decided was that a current tool [CADAM] wasn't good enough for us to compete in this sector of the market, therefore we have got to have shorter design cycles, better design cycles and more effective design cycles, and we felt that not only did we have to look at the CAD system in this respect, but in fact the whole company organisation had to be geared towards these objectives'.

Factors Affecting the Decision

The decision is driven by meeting customer requirements, that the system should be: (1) flexible; capable of doing different sorts of designs, and (2) responsive; so that customers' requirements could be met in a short time scale.

'We are effectively having a complete design cycle for each ship instead of a design cycle which may last for two or three ships. So we have to shorten our own internal system as far as possible'.

The total estimated cost of the project is about £10½ million and should lead to savings in cost and increased turnover. Both of these are being reflected in the financial analysis of the project:

'... The reason for doing it [the replacement] is to make the design more effective in relation to the marketplace. We are obviously trying to do things more cost effectively and we would see savings in a number of areas, clearly in production down-time through to re-work. We see savings in technical design and man-hours, where revisions are not required so readily because of mistakes. We are able to move into production quicker, thus increasing our turnover over a period of time ... The savings and turnover is always a difficult one when you are discussing with the board. What you start to talk about is opportunity costs or opportunity advantages. Essentially, in our case, what we are saying is that if we can produce two ships a year with the current design process and if we can move that to 2.3 ships a year with a new design process, then ostensibly we are increasing our turnover and we will just take the profit on turnover if you like, and use that. But it is based on an increased throughput per year'.

The company uses payback and discounted cash flow methods in evaluating investment projects. Both NPV and IRR are calculated and the focus on one of them depends on the suitability of their results.

'... We will use the one which suits us best...One has to realise that when you are putting these things together that the margin for error is so significant in terms of estimated costs and so on, that you could get hung up on half a percent on DCF. We take a fairly pragmatic view, we will come at it two or three different ways and see where the numbers are focusing, and then we will take a sensible view'.

The company may use different financial techniques when evaluating each investment project, although there is a manual which sets out the basis for how the financial techniques should be applied.

'But at the same time, if we are putting a decision together I would want to satisfy myself that we have our confidence of agreement coming through at us through different ways'.

DCF is theoretically sound but in practice 'it is flawed in relation to AMT particularly ...' DCF is based on assumptions but 'people tend to forget that you have made these significant assumptions... No one in the board would really look at them.' For example, a mainframe investment that cost the company about £2 million fifteen years ago and it is worth today only £15,000.

One risk that may have an effect on AMT decision making is the rapid obsolescence of the technology. Delay means loss of competitive advantage for the company. Projected obsolescence affects: 'your capital write-off policies. For example, you might write-off a PC in six months ... Someone else might say will have it for 10 years... I think that would be a bit risky.'

The expected risk of a project is usually evaluated through sensitivity analysis. 'What if' questions are investigated for only one change at time. No attempt is made to attribute probabilities to the estimates.

> '... You have to understand that from start to finish there is uncertainty in this whole thing. So we would take a pragmatic view based on the projected savings. We may decide to take only 60% of the savings and see how the figures stack back at that kind of level... for example, if I take savings in production re-work. Obviously I will look at the level of re-work just now and I will say, well, where will it be. I will ask a number of people, but none of these people actually know. They will give a view. Somebody will say well current re-work is 6%, we reckon it will go to 4%. Someone else says I think it will go down to 2%. We don't know, that is the realisation ... There is a bit which says OK, these are all focusing at around 3½%, so lets run with that. It sounds a reasonable level so that's the level of saving which we will take. Then we will look at the risk on that, what happens if it was 2%, how would the figures stack'.

The expected return on the project is the most important factor in the final decision. The target rate depends on the investment type. If it is a design tool, like the current project, a 10% return would be acceptable.

Most of the benefits of the project are quantified as financial consequences.

> '... Perhaps it is quite strange to call it non-financial criteria because, from my perspective, all those things are quantifiable to a degree. One has to accept that they are not perfect assumptions, but they are quantifiable... So we have got five or six categories where we would expect to generate savings

in efficiency and quality and re-work, but essentially, they are all financial as it were, because they have a bottom line'.

'... If all I was going to do was add a £1 million or £500,000 into the cost of the operation to meet a marketing imperative, then I would be most reluctant to do it. I am hoping that we might be meeting the marketing imperative, but at the same time, we would seek to make the operation more efficient in other ways'.

Everything related to the project should be quantified in financial terms.

'Unless I could say, by how much and what impact that will have on the business, it's just a motherhood statement. ... It may be helpful to say to somebody as a result of this, we will be able to sustain the employment in the design offices or cope with more change in circumstances, that might be nice to know. Frankly, these are not what the decisions are taken on, people don't take the decision to spend money based on motherhood statements'.

For example, an improvement in quality has a financial consequence.

'So if someone said to me in a capital request, "please sign this". With a statement like that I would say "well how much are we going to save?" Also, something like employee morale and company image may be helpful, but to put forward that, for example, as the prime basis would never work... Any warm feelings that you can give those who are making the decisions is good, but I regard that simply as a warm feeling'.

Formal Process and Presentation

The proposal is prepared in draft and presented at the board as an outline project. Then the final proposal is prepared, discussed locally and finally presented at a board meeting. The decision concerning replacement of the CAD computer system will be taken by the board as its value is large (£10½ million).

The departments which had participated in evaluating the project include engineering, production, finance. Marketing had also participated to the extent that they have debated the strategy and considered: 'why are we making the move?'

The investment decision in this company is summarised in figure 5.7.

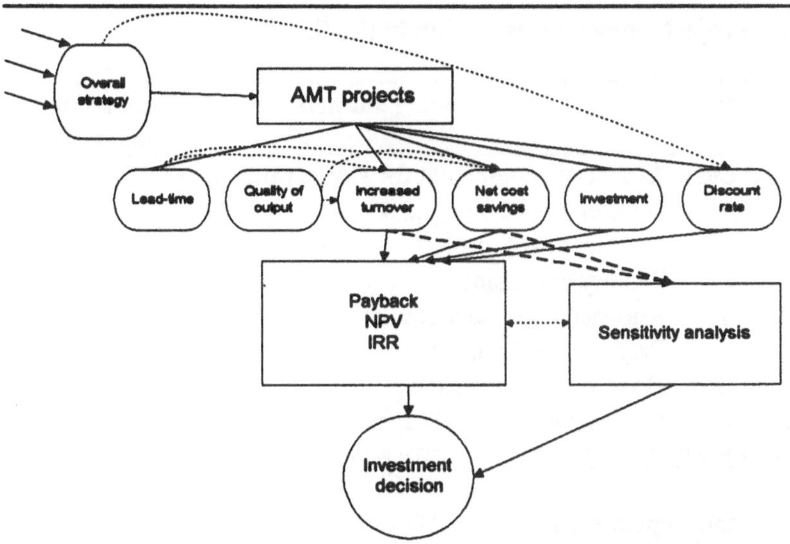

Figure 5.7 The investment decision model in Maritime Ltd

Discussion and Conclusions

The aim of studying a set of cases in AMT investment decision making was to elicit the key factors considered in the decision, and consider how these factors could be reflected in the decision model. Reporting these factors can also help decision analysts to focus on them and assess their importance. Consistent with the survey results reported in the previous chapter and with the capital budgeting literature, it was found that such factors could be categorised into three groups: financial, non-financial and risk factors. In each case, a major recent investment project was taken as an example to illustrate how the decision was taken. However, interviewees were not restricted to this project, when normal practice was different to that in the project considered, they were asked to point this out.

Table 5.1 shows the main projects described in the field study. These projects range from less than a half million pounds to £200 million, and most had been undertaken in the last few years. However, the final two stages of the 'factory expansion' project in Tobacco Ltd had not been

Table 5.1 Projects under investigation in the field study

Company	Project being studied	Total amount invested
Baby Food Ltd	Automated equipment for sleeve printing	£1.6-1.7m
Engineering Plc	Developing a casting process	£16m
Newspaper Plc	Printing direct to plate	£1.4m
Tobacco Ltd	Factory expansion	£200m
Packaging Plc	Knowledge based systems	£450k
Heavy	Product development	£15m
Engineering Plc	Upgrade machine tools	£4.5m
Maritime Ltd	Replacement of CAD system*	£10.5m

* This project was under evaluation at the interview time.

completed, and 'replacement of the CAD system' in Maritime Ltd was in the process of evaluation.

Motivation to Invest in Advanced Technology

Investment in new manufacturing technology is motivated by various causes. Table (5.2) provides a summary of the main reasons for AMT investment reported by the interviewees. It is noticeable that cost savings and improvements of output quality were considered as major reasons in all the cases. Although reduced lead-times is often considered to be one of the main

Table 5.2 Main reasons for investing in AMT

Reasons	A	B	C	D	E	F	G
Cost savings	✓	✓	✓	✓	✓	✓	✓
Improve output quality	✓	✓	✓	✓	✓	✓	✓
Reduce lead-times		✓				✓	✓
Secure a competitive position		✓		✓		✓	
Quick response to customer needs			✓		✓		✓
Increase turnover					✓		✓
Reduce possibility of breaking down	✓						

A: Baby Food Ltd, B: Engineering Plc, C: Newspaper Plc, D: Tobacco Ltd, E: Packaging Plc, F: Heavy Engineering Plc, G: Maritime Ltd.

motivations for investing in AMT according to the literature, in only three cases was it considered as one of the main reasons. This might be due to the nature of manufacturing in two of the companies that specialise in continuous processes. Most interviewees did not consider an increase in turnover as one of the main reasons for investing in AMT, rather, cost savings were the usual financial reason in the preliminary justification of an AMT investment.

Financial Return Measures

The expected financial return from an investment project is the most important factor in the decision. Interviewees confirmed their use of the four well-known financial methods for evaluating AMT investments (see table 5.3). In the case of Packaging Plc the impact on the return on sales (ROS) was one of the most important parameters in the evaluation of investment projects. This measure was justified by the interviewee because ROS is used in the company as a means of evaluating performance and rewarding managers. At least two financial methods were used in the evaluation of investment projects in all the companies. Many interviewees justified their use of more than one financial measure because it provides the decision maker(s) with more information and/or helps them to cope with the different preferences of multiple participants in the decision making process. However, the finance director of Maritime Ltd confessed that the calculation of both NPV and IRR were undertaken in order to select the one which would be most helpful in gaining the approval of the board.

Payback is the method used in all companies, and this together with the survey results confirms the practical importance of this method over all others (see also for example: Pike, 1996; Carr et al., 1994; Lefley, 1994a). Accordingly, investments should show short-term results in order to be approved. In Heavy Engineering Plc., a distinction is drawn between long- and short-term projects in assessing the importance of the financial measures used, but it was surprising that ROI was considered to be a long term measure.

Non-financial Criteria

It was concluded from the questionnaire responses that non-financial criteria play an important role in the final decision for both AMT and for other investments. In the questionnaire, a number of non-financial criteria, usually

Table 5.3 Financial appraisal methods used

Method	Baby Food Ltd[a]	Engineering Plc[b]	Newspaper Plc	Tobacco Ltd[c]	Packaging Plc[d]	Heavy Engineering Plc[e]	Maritime Ltd
Payback	Yes	Yes	Yes	Yes[i]	Yes	Yes	Yes
NPV	Limited	Yes[g]	No	Yes	Limited	No	Yes[j]
IRR	Yes[f]	Limited	No	No	No	No	Yes[j]
ROI	No	No	Yes[h]	Yes	Yes	Yes	No

[a] Projects with a payback of 5 years or more should be excluded from any further analysis.
[b] Payback is the main criterion for simple projects while NPV is used only for major projects.
[c] Payback used as a measure of liquidity.
[d] The impact of the investment on the return on sales was used as a key parameter for evaluating the investment project.
[e] Payback is the main criterion in approving short-term projects (at least a period of three years is required) while ROI is the main criterion in approving long-term projects (at least a minimum required rate is 25%).
[f] 22% is the minimum required rate.
[g] A payback of three years or less is required for approving a project.
[h] 20% is the minimum required rate.
[i] The calculation is based on discounted cash flows.
[j] Either NPV or IRR is reported for the project and the choice of one of them depends on the suitability of the results for obtaining the approval of head office.

Table 5.4 The non-financial criteria

	A	B	C	D	E	F	G
Overall strategy consideration	✓	✓	✓	✓	✓	✓	✓
Quality of the output	✓	✓	✓	✓	✓	✓	✓
Customer requirements		✓	✓	✓	✓		
Obtaining greater flexibility	✓			✓		✓	
Lead-times		✓				✓	✓
Keeping-up with competition		✓			✓		
Employees working conditions		✓					
Track record		✓			✓		
Limited space				✓			

A: Baby Food Ltd, B: Engineering Plc, C: Newspaper Plc, D: Tobacco Ltd, E: Packaging Plc, F: Heavy Engineering Plc, G: Maritime Ltd.

cited in the literature as important factors in the investment decision, were listed and respondents were asked to rate each criterion using a five-point equal distance scale. In the interviews, a different approach was followed; interviewees were asked to indicate the important factors which affect investment decisions but were not prompted by identifying specific criteria. In this way, it was believed that interviewees would discuss only the most important factors relating to the case study decisions.

The importance of non-financial aspects in the investment decision analysis has been extensively discussed in recent research such as Carr and Tomkins (1996), Tomkins and Carr (1996), Van Cauwenberg et al. (1996), Shank (1996), and Slagmulder (1997). For example, Shank (1996) used a case study to demonstrate the limitations of NPV in investment decisions. He argued that NPV needs to be complemented by the use three strategic tools: value chain, cost driver, and competitive advantage analyses. Although traditional financial measures of an investment project indicated a very favourable return, Shank's strategic analysis for the same project tended to reverse the decision.

From the cases reported, it can be seen how different non-financial issues may affect the investment decision. Table 5.4 summarises the non-financial factors discussed in each case. In general, although in most cases interviewees explained the importance of these non-financial factors, there is little evidence that they were considered in a formal way. In three cases (Newspaper Plc, Tobacco Ltd and Packaging Plc) limited use was made of scoring methods. In Newspaper Plc. the scoring method was used in the choice between the suppliers, in Tobacco Ltd it was used for some environmental investment decisions, and in Packaging Plc it was only used by the American subsidiaries.

In many cases the discussion of non-financial factors in the evaluation process is used to supplement the financial results. One exception to this is the strategic consideration of an investment. In almost all the cases, investments were reviewed for conformity with overall strategy before undertaking further detailed analyses. If an investment did not conform with the overall strategy, then it simply would not be considered in detail. The importance of the consistency of a new investment with the overall strategy confirms the earlier findings of all the empirical studies that considered this issue; such as Slagmulder and Bruggeman (1992), Carr et al. (1994), and Nixon (1995). Slagmulder's (1997) study revealed that failure to ensure the consistency of a new investment with overall strategy can lead to strategically beneficial projects being overlooked or, conversely, the approval of projects that should not have been subjected to detailed analysis.

The quality of output represents an over-riding issue in most cases. This may be because all the companies studied are large and most are leaders in their industries. Many of the interviewees considered that the quality of output criterion has both financial and non-financial implications. For example, improving the quality of output may result in cost savings in areas such as inspection, scrap and rework, as well as increasing turnover. In addition, the improvement of output quality may affect the competitive position of the firm, satisfaction of customers and reduce lead-times.

The dominance of financial considerations was clear in two cases, Newspaper Plc and Maritime Ltd. In these cases the interviewees believed that if there is a benefit related to the investment project, it should be quantified in financial terms. One of these interviewees described the non-quantifiable benefits as a 'warm feeling' that supported the decision, but that these non-quantifiable benefits would not make any difference in deciding whether to accept an investment project. Accordingly, the investment project would be evaluated only on the basis of payback and discounted cash flow.

Risk Analysis

The questionnaire responses in relation to risk analysis did not give a sufficiently clear idea about how practitioners dealt with risk in investment projects. This issue was further investigated and reported in the seven field study cases.

The interviewees confirmed that sensitivity analysis was routinely carried out and was given considerable attention during the decision making process. Only in very few cases was formal consideration given to the possibility of ruinous loss or liquidity implications, and project variability was addressed in a relatively unsophisticated manner through sensitivity analysis.

During discussions with the interviewees, it was clear that they were more interested in minimising the risk of investment projects than measuring it. Risk was usually analysed in terms of its sources and of how it might be eliminated or minimised. In general, two types of risk were illustrated by the interviewees: risk related to the market and risk related to the technology. The project risk related to the market might arise from the volatility of expected demand and the life of the market. The project risk related to technology might arise from the failure of suppliers to meet the specifications and/or the possibility of choosing the wrong technology. Each type of risk could be reduced, eliminated or shared with other parties, such as customers and suppliers. For example, the volatility of expected demand could be reduced through commissioning a survey to forecast the likely demand of proposed products. In the case of an investment project related to the introduction of new products, it might be wise to introduce them in stages rather than simultaneously. The risk of supplier failure could be reduced through piloting the new technology. In addition to these two main types of risk (market and technology), some interviewees added liquidity risk, suggesting that this type of risk could be dealt with by accepting only projects with short paybacks.

In summary, it seems that practitioners use sensitivity analysis in order to identify key variables, and then take action in order to minimise any obvious risk regarding these variables. The implicit model used by many practitioners seems to incorporate residual risk, which cannot be shared with other parties, by using high discount rates and short payback periods.

It can be concluded that practitioners combine both moment-oriented and dimension-oriented approaches when evaluating project risk. First, the dimension-oriented approach is employed by analysing project risk into its sources (dimensions), and then effort is exerted to minimise the risk associated with each source. Finally, the moment-oriented approach is employed by adjusting the discount/target rate.

A General AMT Investment Decision Model

It is evident from the cases reported that practitioners employ models in AMT investment decision making which, either implicitly or explicitly, combine financial return, risk and intangible benefits. While figures 5.1 to 5.7 summarise the AMT investment decision making model in each company in the field study, figure 5.8 is an attempt to synthesise all the individual models into a general AMT investment decision model.

Figure 5.8 synthesises the main results of the empirical study. The investment decision is affected by several factors; some of which can be easily quantified in financial terms while other factors are difficult to quantify. In principle, as discussed earlier, these factors could be analysed into three dimensions: financial factors, non-financial factors and risk factors. The financial factors, which include the investment amount, net cost savings[7], increased turnover and the discount rate, can be analysed using traditional financial measures, such as payback, ROI and discounted cash flow methods. This group of factors may represent the most crucial considerations in the final decision. Indeed, in some the cases, the decision was totally based on financial analysis together with the results of sensitivity analysis (Newspaper Plc and Maritime Ltd). In most of the cases at least one of the discounted methods was used together with the payback method. Only in Newspaper Plc. and Heavy Engineering Plc was ROI used with the payback method and in Packaging Plc. NPV was used on a limited basis. In the latter company the impact of the investment on return on sales was the main criterion when evaluating and approving the investments.

Practitioners consider non-financial factors, which include intangible benefits, in AMT investment project appraisal. Consistent with the questionnaire findings, the interviewees acknowledged the importance of such factors in the decision, but none of them demonstrated any formal analysis of such factors. They said that, in principle, an AMT investment project should show a potential financial benefit if it is to be accepted. In many cases, non-financial factors may be discussed after ensuring the financial viability of the project. However, the importance of such discussion is that it may lead to approval of projects with lower financial returns because the projects have non-financial benefits which cannot be evaluated in financial terms. In addition, sometimes a project may be excluded from consideration because of a non-financial aspect. For example, in Tobacco Ltd a project was rejected because it would not be sufficiently flexible as regards expected changes in products, even though it showed a good

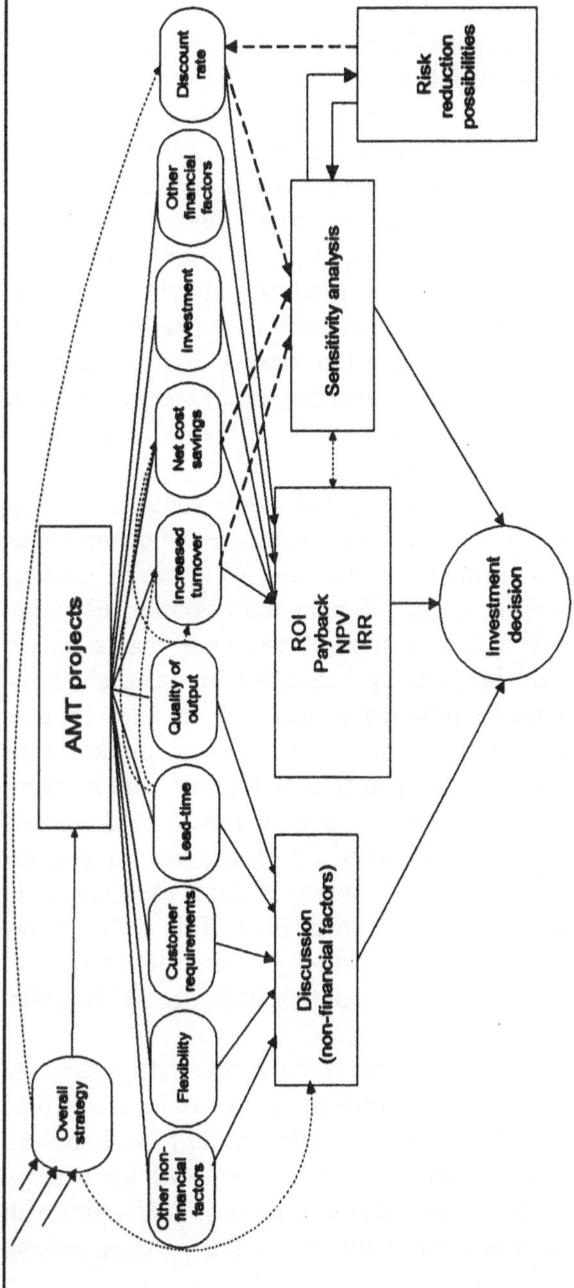

Figure 5.8 A general model for AMT investment decision making

projected financial return. The discussion of intangibles together with the financial return of a project may also reveal biases in the estimates (Nixon, 1995).

Issues such as the improvement of product quality, meeting customer requirements, increased manufacturing flexibility, conformity with overall strategy and lower inventory/reduced lead times represent the most important non-financial factors. While most interviewees recognised that this group of factors represents an important issue to be considered in the final decision[8], some interviewees (especially in Newspaper Plc and Maritime Ltd) said that such factors could only act as a 'warm feeling'. However, most of the interviewees noted that the improvement of product quality has financial consequences as well as non-financial implications. Improvements in product quality usually result in reduced costs of scrap, rework and inspection in addition to the potential increase in sales. It also may lead to improvement of the company image and of the competitive position of the company, which are both hard to quantify in financial terms.

Reduction in inventory levels is most appropriately considered in the financial part of the decision, because such reductions can be quantified as savings resulting from the reductions in the cost of holding, moving, scheduling and inspecting inventories (see, Kaplan and Atkinson, 1989, pp. 485-486). Many of the interviewees expressed the view that expected reductions in lead-times could be partially quantified in financial terms, because such reductions could be reflected in increases in sales volume. However, there may be other benefits from the reduction in the lead-time which would be more difficult to quantify in financial terms, such as faster response to customer requirements. In this respect, it should be noted that reductions in inventory levels and lead-times are closely related because reduced inventory will almost certainly induce reduced lead-times in processing products from raw material to completion. This is intuitively obvious if one considers that existing work-in-process inventory will normally be processed before a new job or batch (Abdel-Kader and Dugdale, 1996).

The conformity of the investment project with overall strategy seems to be an over-riding issue in most companies (indeed all companies in the field study took this view). Any AMT project should conform with overall strategy. If a proposed investment project did not conform with the strategy it, simply would not be considered. According to this view, conformity with strategy hardly needs to be considered explicitly in the decision making model, because all the alternatives under consideration will be congruent with broad company plans.

It should be noted that overall strategy plays an important role in determining the discount/hurdle rate used in the discounting methods. Several interviewees indicated that the rate is determined by head office, is rarely changed, and is not subject to changes when sensitivity analysis is undertaken. Further, none of the interviewees could give a clear rationale for the choice for their particular discount rate; simply that it was just the average of the industry. Although most interviewees in companies that used discounting methods said that the rate is risk adjusted; no change was made to reflect the expected different levels of project risk. However, this implies that all projects are equally risky.

The figures 5.1 to 5.7 indicate that sensitivity analysis is the most used formal method of analysing project risk. As discussed earlier, project risk is usually analysed in terms of its sources and then, where possible, risks are eliminated or shared with other parties, such as suppliers, who may be more capable of dealing with certain types of risk. Thus sensitivity analysis may play an important role in determing key sensitive factors which are then further analysed for possibilities of reduction or elimination. Clearly, not all risks can be eliminated or reduced because uncertainties always remain. Sensitivity analysis represents a way of providing the decision maker(s) with a way of understanding the effects of changes in key variables on the potential profitability of an investment project. Variables that may be subject to sensitivity analysis include net cost savings and increased turnover. The discount rate may also be considered.

Figure 5.8 illustrates a general model for AMT investment decisions derived from a synthesis of the cases in the field study. This model provides the basis for a suggested AMT investment model utilising the mathematics of fuzzy set theory. Fuzzy set theory is introduced in the next chapter, and the suggested model is developed in chapter 7.

Notes

1 The content and presentation of each case was approved by the interviewee.
2 All company names are pseudonyms.
3 These cases resulted from five face-to-face interviews and two telephone interviews.

4 In July, 1996 (2 months after the interview was held), the company had a new CEO and a new chairman who decided to sectorise the business into seven areas. Four areas are in packaging: food and beverage, beauty, health care, and industrial. The other three areas are in speciality films and papers, printing, and building and engineering. Each sector is expected to run its own innovation and technology development.

5 When the case was sent to the interviewee in order to review and check it against company practice, he stated that the company had changed the key financial criterion back to return on net assets. This change was aimed at increasing the assets productivity and minimising 'technological' expenditure.

6 The system is not normally used outside the shipbuilding and aerospace sectors.

7 The term net cost savings is used to refer to the expected total cost saving less any expected increases in costs resulting from operating the new automated manufacturing systems.

8 The analysis of the questionnaire results reported in the previous chapter confirmed such a view in that a significant difference between AMT investment projects and other types of investment projects was found in factors of 'improving product quality', 'increased manufacturing flexibility', 'reduction in inventory levels' and 'reduced lead times'.

6 Fuzzy Set Theory

Chapter Preview

This chapter presents the mathematics of fuzzy set theory in order to provide a basis for developing a model for AMT investment decision making. Although some readers may find this chapter difficult, it is not necessary to understand all the detailed mathematical analyses in order to follow the main arguments. Less mathematically inclined readers should read the introductory section (*Basics of Fuzzy Sets*) to gain a general understanding of what is meant by a fuzzy set. This section introduces the mathematical definition of a fuzzy set (equation 6.1) and the meaning of this can be intuitively grasped through the examples given (A, B and C to represent low, medium and high production output). A further example, of a real number 'close to 10' is illustrated and , again, this should help the reader to clarify what is meant by a 'fuzzy set'. Two sub-sections (*Basic Fuzzy Set Concepts* and *The Main Operations on Fuzzy Sets*) introduce some basic concepts that underpin the mathematics of fuzzy sets.

The next section (*Mathematics of Fuzzy Numbers*) develops the fuzzy mathematical operations that will be needed in the development of the AMT decision making model. Much of this may be skimmed, but it is suggested that figures 6.7, 6.8 and 6.9 be studied so that the ideas of fuzzy mathematics are grasped.

The sub-section '*Special Fuzzy Numbers*' introduces certain special fuzzy numbers that have been suggested in the literature: L-R fuzzy numbers, L-R flat fuzzy numbers, triangular fuzzy numbers and trapezoidal fuzzy numbers. Here, again, non-mathematically inclined readers might find the analysis difficult but they should, nevertheless, be able to grasp the basic idea behind each of the special types from figures 6.10, 6.11, 6.12 and 6.13, as these are important in chapter 7.

The problem of ranking fuzzy numbers is introduced in the next section (*Ranking Fuzzy Numbers*) and different methods of ranking suggested in the literature are reviewed. In order to understand the ranking method proposed later in the chapter it is first necessary to understand the method suggested by Jain and outlined in the subsection '*The Maximising Set Approach*'. The idea can be intuitively grasped by reference to figure 6.15 which ranks three fuzzy numbers A, B and C, as C greater than B and B as greater than A. the problem with Jain's method is that it ranks the numbers only by reference to the right-hand-sides of their membership functions. Chen suggested that this deficiency could be overcome by a method that took account of both the right-hand and left-hand sides of the fuzzy numbers to be ranked. Chen's ideas are developed in the subsequent sub-section (*Minimising and Maximising approach*) together with criticisms of Chen's method and its further development by Chen and Hwang.

This is followed by the introduction of a new method for ranking fuzzy numbers that, in the opinion of the authors, is intuitively attractive, takes account of *all* the information available in the fuzzy numbers under consideration, and allows for different rankings dependent on the 'optimism' or 'pessimism' of the decision maker(s).

The next section '*Linguistic Variables*' deals with the ideas behind fuzzy linguistic variables. This section is easier than the others and so it should not be difficult to grasp the idea of a fuzzy linguistic scale and the alternative scales suggested by Chen et al. that are illustrated in figures 6.25a-6.25h.

The last section, *Conclusions*, provides an overview and summary of the chapter as a whole.

Introduction

A decision might be affected by one or more of three types of variable: quantitative, qualitative, and approximately quantitative. The definition of these types is based on the nature of the values assigned to them. Quantitative variables are assigned crisp numbers whereas qualitative variables take verbal values such as 'high', 'low', 'fast', 'slow', etc. The third type: approximately quantitative variables, might be assigned values such as 'around £300', 'between £4000 and £5000', 'approximately 10%', etc. Both qualitative and approximately quantitative variables are difficult to express in precise numerical values; they are *fuzzy* variables.

In 1965 Zadeh proposed fuzzy set theory, a set of mathematical concepts and techniques for modelling fuzzy variables. The term 'fuzzy' refers to the situation where ambiguity and vagueness exist[1]. In these situations, traditional two value, true and false, logic breaks down, and instead, there may be many values between these two extremes (Klir and Folger, 1988). Fuzzy logic addresses these situations by allowing variables to be 'partially true' and/or 'partially false'. Fuzziness must be differentiated from randomness. According to Bellman and Zadeh (1970) randomness involves uncertainty, expressed as the probability of occurrence or non-occurrence of an event, whereas fuzziness deals with the imprecision and inexactness of words, concepts, events, and judgements. For example, the expression 'massive output' involves ambiguity due to the imprecise meaning of the word 'massive'. However, a question about the probability of producing 1000 units involves randomness. The event is well described, either the production is 1000 units or it is not, the uncertainty lies in the occurrence or non-occurrence of the event (Zebda, 1989). However, randomness, as measured by probability, can be related to fuzzy events. For example, the probability of having massive output might be 0.8. On the other hand, judgements about probability can be vague or ambiguous, for example, the probability of having production of 1000 units could be 'about 80%' (Zebda, 1989). In short, while fuzziness deals with uncertainty in the meaning of an event, word, or concept, randomness deals with the uncertainty of occurrence or non-occurrence of the event itself. Randomness is analysed using probability theory while fuzziness is analysed using fuzzy set theory.

The aim of this chapter is to introduce the framework of fuzzy set theory in order to utilise it in the formulation of a model for evaluating AMT investments[2]. Much of this framework is based on existing fuzzy set theory literature. However, an important contribution is made in this chapter: a novel method for ranking fuzzy numbers[3]. This proposed ranking method will be used in ranking AMT investment projects in the AMT decision model developed in chapter 7.

The remainder of this chapter is divided into five sections as follows:

1. Basics of fuzzy sets.
2. Mathematics of fuzzy numbers.
3. Ranking fuzzy numbers.
4. Linguistic variables.
5. Conclusions.

Basics of Fuzzy Sets

A set, in general, is a collection of objects or elements which have a common or a specific feature. The set might be finite, such as a set of 'all first year students in the university', or infinite such as the set of 'real numbers'. According to Zimmermann (1991, p. 11) sets can be described in one of three forms. Firstly, a set can be described by listing all of its elements, for example A= {a, b, c, d}. Secondly, a set can be described analytically be stating the condition for membership, for example $A = \{x \mid 5 < x \leq 10\}$. Finally, a characteristic function may be used to define the member elements. In classical set theory, the characteristic function is either 1 or 0; 1 indicates that an element belongs to the set, while 0 indicates that the element does not belong to the set[4]. On the other hand, in fuzzy set theory, the characteristic function allows various degrees of membership (or belonging) of the members of a specific set.

Zadeh (1965) in his early paper defined a fuzzy set as 'a class of objects with a continuum of grades of membership'. In other words, the set is described as fuzzy if there is no sharp boundary between those elements (or objects) which belong to the class and those which do not (Bellman and Zadeh, 1970). For example, the set of 'large companies' is a fuzzy set since there is no sharp boundary between what is large and what is not. Also, classes of 'beautiful women', 'young men', 'red flowers', 'small cars', etc. are all fuzzy sets (Zadeh, 1972).

Mathematically, let $U = \{ x_1 \ldots\ldots\ldots x_n\}$ be a universal set of objects x, then a fuzzy set A in U is defined by ordered pairs[5]:

$$A = \{ x, \mu_A(x) \}, \qquad \forall x \in U \tag{6.1}$$

where $\mu_A(x)$ is called the membership grade of x in A, and $\mu_A: U \rightarrow M$ is a function from U to a space M. Usually M takes a real number in the closed interval [0,1] where 0 and 1 refer to full membership and non-membership respectively. When M contains only two points, 0 and 1, A is non-fuzzy and its membership function is identical to the characteristic function of a classical (crisp) set (Bellman and Zadeh, 1970).

The membership function indicates the extent to which each element (or object) in the universal set, U, belongs to the fuzzy set, A. In the fuzzy set theory literature, the membership function is referred to as 'degree of compatibility', 'grade of membership', or 'degree of truth'. As an example to indicate the fuzzy set and its membership function, let U = {10, 20, 30, ... 100} be a set of possible daily production output. Then, fuzzy sets A, B, C in U can be defined (subjectively) as follows:

A = 'Low production output'
= {(10, 1), (20, .8), (30, .5), (40, .2)}
B = 'Medium production output'
= {(30, .7), (40, 1), (50, 1), (60, .8)}
C = 'High production output'
= {(50, .2), (60, .6), (70, .8), (80, 1), (90,1), (100,1)}

Referring to the fuzzy set, A, it can be seen that a production of 10 units is assigned a membership function '1' because it is considered absolutely as a low production level, but, on the other hand, production of 20 units is assigned a membership function of less than '1' as it is considered to be 'more-or-less' a low production level. Production levels of 50, 60, ..., 100 are assigned membership function, 'zero' because they are not considered as low production levels at all, and as here, such elements are normally not listed (Zimmermann, 1991, p. 12).

A fuzzy set can be described in two forms, other than (6.1) (Zadeh, 1973). Firstly, if U is a finite set $\{x_1, \ldots\ldots x_n\}$, then a fuzzy set A can be described as:

$$A = x_1/\mu_A(x_1) + \ldots\ldots\ldots\ldots x_n/\mu_A(x_n) \quad (6.2)$$

or

$$A = \sum_{i=1}^{n} x_i|\mu_A(x_i) \quad (6.3)$$

Note that, here + and Σ do not mean addition. They are used to represent a collection of elements belonging to a fuzzy set.

Secondly, if U is an infinite set, then a fuzzy set A might be written as:

$$A = \int_X x|\mu_A(x) \quad (6.4)$$

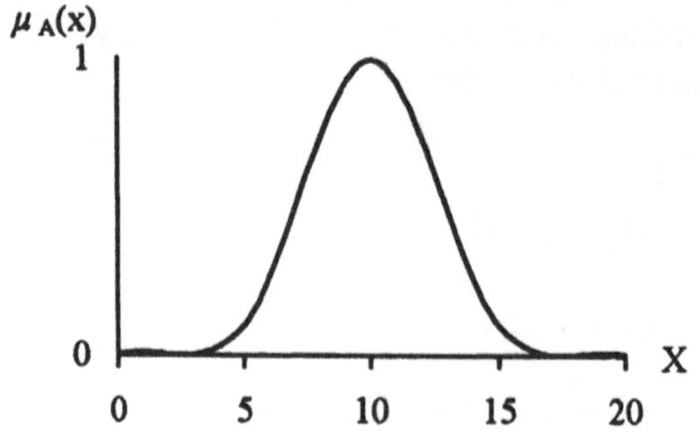

Figure 6.1 Real number close to 10

Source: Zimmermann, 1991, p. 13.

Note that ∫ does not refer to the integration sign. It is used to represent infinite elements belonging to a fuzzy set.

For example, the fuzzy set A = 'real numbers close to 10' may be represented as in figure 6.1 (Zimmermann, 1991, p. 13) and by the equation:

$$A = \int_{\Re} x \,\Big|\, \frac{1}{1+(x-10)^2} \tag{6.5}$$

where $\Re$ is the real line.

Basic Fuzzy Set Concepts

The definitions of the most common concepts in fuzzy set literature are described below (Bellman and Zadeh, 1970; Zimmermann, 1991; Klir and Folger, 1988; and Dubois and Prade, 1980).

The Support of a Fuzzy Set The support of fuzzy set A, S(A), is the crisp set of all elements which have membership grade more than 0. That is:

$$S(A) = \{x \in U \mid \mu_A(x) > 0\} \tag{6.6}$$

For example, the support of the fuzzy set 'low production output', in the previous example, is:

$$S(A) = \{10, 20, 30, 40\}$$

all of these x values have corresponding $\mu_A(x) > 0$.

α- Level Set (α- Cut) of a Fuzzy Set An α- cut of fuzzy set A is a crisp set, A_α, the elements of which belong to A at a level to a defined degree of α. That is:

$$A_\alpha = \{x \in U \mid \mu_A(x) \geq \alpha\} \tag{6.7}$$

The α- cut is a more general case of the support of a fuzzy set. Where α= 0, then $A_\alpha = S(A)$ (Chen et al., 1992, p. 46). Moreover, a strong α- cut is defined as (Zimmermann, 1991, p. 14):

$$A'_\alpha = \{x \in U \mid \mu_A(x) > \alpha\} \tag{6.8}$$

For example, the possible α- level sets of the fuzzy set 'low production output' are:

$$A_1 = \{10\},$$
$$A_{.8} = \{10, 20\},$$
$$A_{.5} = \{10, 20, 30\},$$
$$A_{.2} = \{10, 20, 30, 40\}$$

The strong α- level set for α = .5 is:

$$A'_{.5} = \{10, 20\}$$

The Convexity of a Fuzzy Set A fuzzy set A is convex if and only if:

$$\mu_A(\lambda x_1 + (1-\lambda)x_2) \geq \min\,[\mu_A(x_1), \mu_A(x_2)] \quad \forall\, x_1, x_2 \in U, \text{ and } \lambda \in [0,1] \tag{6.9}$$

Figures 6.2 and 6.3 illustrate a convex fuzzy set and a non-convex fuzzy set respectively. Note that the convex fuzzy set includes only one peak. In contrast, the non-convex fuzzy set includes more than one peak.

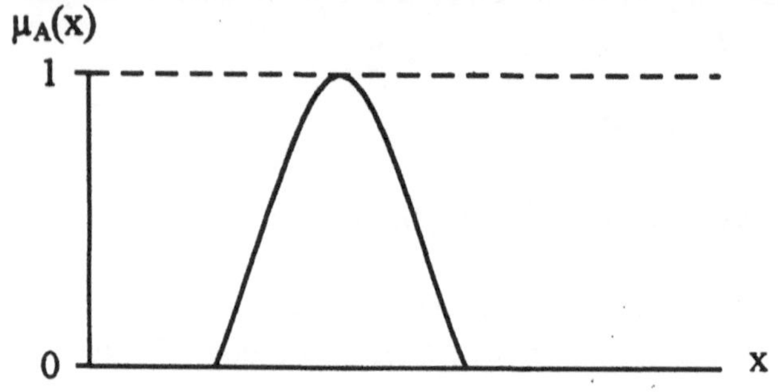

Figure 6.2 A convex fuzzy set

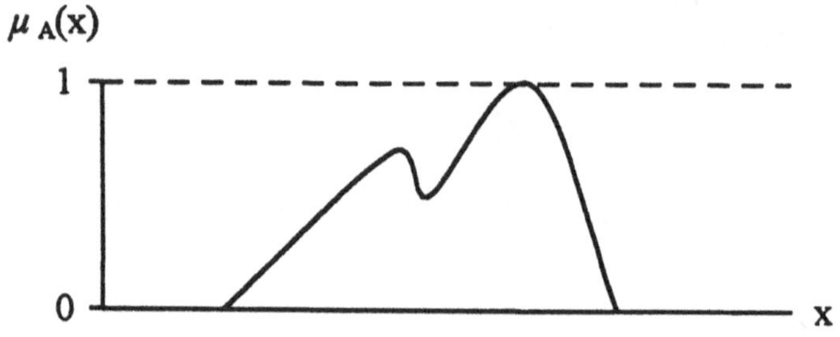

Figure 6.3 A non-convex fuzzy set

The Normality of a Fuzzy Set A fuzzy set A is normal if and only if sup_x $\mu_A(x) = 1$, that is, the supremum, or lowest upper bound, of $\mu_A(x)$ over U is unity. A fuzzy set is subnormal if it is not normal. A non-empty subnormal fuzzy set can be normalised by dividing $\mu_A(x)$ by the $sup_x\ \mu_A(x)$, (Bellman and Zadeh, 1970). The sets in figures 6.2 and 6.3 are normal fuzzy sets. The normality of a fuzzy set guarantees that at least one element fully satisfies the phenomenon to which the fuzzy set applies (Chen et al., 1992, p. 48).

The Cardinality (or Power) of a Fuzzy Set The cardinality of fuzzy set A evaluates the proportion of elements of U having the property A (Chen et al., 1992, p. 48). For a finite fuzzy set A, the cardinality | A | is defined as:

$$|A| = \sum_{x \in U} \mu_A(x) \tag{6.10}$$

And, the relative cardinality of A is defined as:

$$\|A\| = \frac{|A|}{|U|} \tag{6.11}$$

The relative cardinality can be interpreted as the proportion of elements of U being in A, weighted by their membership function in A (Zimmermann, 1991, p. 16). For example, the cardinality and relative cardinality of the fuzzy set 'low production output' are computed as:

$$|A| = 1.0 + 0.8 + 0.5 + 0.2 = 2.5$$
$$\|A\| = 2.5/10.6 = 0.24$$

The Main Operations on Fuzzy Sets

The main operations on ordinary sets: complement, intersection, and union, are extended to fuzzy sets by using their membership function as follows:

1. The complement of a fuzzy set A, CA, is defined as:

$$CA = \{x, \mu_{CA}(x)\} \qquad \forall x \in U \tag{6.12}$$

where:

$$\mu_{CA}(x) = 1 - \mu_A(x) \tag{6.13}$$

2. The intersection between two fuzzy sets, $A \cap B$, is defined as:

$$A \cap B = \{x, \mu_{A \cap B}(x)\} \qquad \forall x \in U \tag{6.14}$$

where:

$$\mu_{A \cap B}(x) = \wedge\, [\mu A(x)\, ,\, \mu B(x)] \tag{6.15}$$

where: $\wedge$ denotes minimum value.

3. The union of two fuzzy sets, $A \cup B$, is defined as:

$$A \cup B = \{\, x,\, \mu_{A \cup B}(x)\} \qquad \forall\, x \in U \tag{6.16}$$

where:

$$\mu_{A \cup B}(x) = \vee\, [\, \mu_A(x)\, ,\, \mu_B(x)] \tag{6.17}$$

where: $\vee$ denotes maximum value.

For example, let $U = \{x_1, x_2, x_3, x_4, x_5\}$ be the available alternative production systems, and A and B are fuzzy sets of U referring to the sets of flexible systems and inexpensive systems, respectively. A and B are defined (subjectively) as follows:

$A = \{(x_1, 0.2), (x_2, 0.8), (x_3, 1.0), (x_4, 0), (x_5, 0.5)\}$
$B = \{(x_1, 0.5), (x_2, 0.3), (x_3, 0.2), (x_4, 1.0), (x_5, 0.8)\}$

The membership function of CA, $A \cap B$, and $A \cup B$, are determined using equations 6.12 - 6.17 as shown in table 6.1.

The complement, intersection, and union - as defined above - can be interpreted as the operators (connectives) 'not', 'and', and 'or' respectively (Zebda, 1989). For instance, referring to the above example, CA represents

Table 6.1 Example of complement, intersection and union of two fuzzy sets

U	x_1	x_2	x_3	x_4	x_5
$\mu_A(x)$	0.2	0.8	1.0	0.0	0.5
$\mu_B(x)$	0.5	0.3	0.2	1.0	0.8
$\mu_{CA}(x)$	0.8	0.2	0.0	1.0	0.5
$\mu_{A \cap B}(x)$	0.2	0.3	0.2	0.0	0.5
$\mu_{A \cup B}(x)$	0.5	0.8	1.0	1.0	0.8

a set of the systems which featured as 'not' flexible, $A \cap B$ represents a set of the systems which featured as flexible 'and' inexpensive, $A \cup B$ represents a set of the systems which featured as flexible 'or' inexpensive[6].

Zimmermann (1991, p. 19) reported the following algebraic properties for intersection (= and) and union (= or), assuming A, B, and C are fuzzy sets:

1. Commutativity:

$$\mu_A(x) \wedge \mu_B(x) = \mu_B(x) \wedge \mu_A(x) \tag{6.18}$$
$$\mu_A(x) \vee \mu_B(x) = \mu_B(x) \vee \mu_A(x) \tag{6.19}$$

2. Associativity:

$$(\mu_A(x) \wedge \mu_B(x)) \wedge \mu_C(x) = \mu_A(x) \wedge (\mu_B(x) \wedge \mu_C(x)) \tag{6.20}$$
$$(\mu_A(x) \vee \mu_B(x)) \vee \mu_C(x) = \mu_A(x) \vee (\mu_B(x) \vee \mu_C(x)) \tag{6.21}$$

3. Distributivity:

$$\mu_A(x) \wedge (\mu_B(x) \vee \mu_C(x)) = (\mu_A(x) \wedge \mu_B(x)) \vee (\mu_A(x) \wedge \mu_C(x)) \tag{6.22}$$
$$\mu_A(x) \vee (\mu_B(x) \wedge \mu_C(x)) = (\mu_A(x) \vee \mu_B(x)) \wedge (\mu_A(x) \vee \mu_C(x)) \tag{6.23}$$

4. $\mu_A(x) \wedge \mu_B(x)$ and $\mu_A(x) \vee \mu_B(x)$ are continuous and nondecreasing in each component. Intuitively, $\mu_{A \cap B}$ or $\mu_{A \cup B}$ cannot decrease when $\mu_A(x)$ or $\mu_B(x)$ increases. Further, a small increase of $\mu_A(x)$ or $\mu_B(x)$ cannot induce a strong increase in $\mu_{A \cap B}$ or $\mu_{A \cup B}$.

5. $\mu_A(x) \wedge \mu_B(x)$ and $\mu_A(x) \vee \mu_B(x)$ are strictly increasing in $\mu_A(x)$. If $\mu_A(x_1) = \mu_B(x_1) > \mu_A(x_2) = \mu_B(x_2)$, then, $\mu_{A \cap B}(x_1)$ or $\mu_{A \cup B}(x_1)$ is certainly strictly greater than $\mu_{A \cap B}(x_2)$ or $\mu_{A \cup B}(x_2)$ (Pal and Dutta Majumder, 1987, p. 42).

6. $\mu_A(x) \wedge \mu_B(x) \leq min\ (\mu_A(x), \mu_B(x))$ (6.24)
   $\mu_A(x) \vee \mu_B(x) \geq max\ (\mu_A(x), \mu_B(x))$ (6.25)

This implies that membership in $A \cap B$ requires more, and membership in $A \cup B$ requires less.

7. $1 \wedge 1 = 1$ and $0 \wedge 0 = 0$. This implies that complete membership in A and in B implies complete membership in $A \cap B$, and complete lack of membership in A and B implies complete lack of membership in $A \cup B$.

Mathematics of Fuzzy Numbers

The main aspects of fuzzy numbers are described in this section as follows:

1. The definition of a fuzzy number.
2. The extension principle.
3. Algebraic operations of fuzzy numbers.
4. Special fuzzy numbers.

The Definition of a Fuzzy Number

A fuzzy number is a term used to describe an imprecise quantity. For example, a term such as 'approximately 5' is a fuzzy number, as are, 'about 10', and 'close to 3'. In fuzzy set theory, a fuzzy number 'A' is defined as a fuzzy subset of the real line ($\Re$) which satisfies two conditions, that is, normality and convexity[7] (see for example: Kaufmann and Gupta, 1991; 1985 and Dubois and Prade, 1980). To treat a quantity as a fuzzy number, it is essential that it satisfies both of these conditions simultaneously. For example, let A, B, and C be fuzzy sets in $\Re$ defined as follows:

$$A = \{(7, .4), (8, .6), (9, .8), (10, 1), (11, .7), (12, .5)\}$$
$$B = \{(3, .2), (4, 1), (5, .4), (6, .8), (7, .3)\}$$
$$C = \{(5, .4), (6, .5), (7, .8), (9, .2)\}$$

The fuzzy set A is a fuzzy number (approximately 10) as it is a normal and convex set (see figure 6.4a). On the other hand, neither B nor C are fuzzy numbers because B is a normal but non-convex set (see figure 6.4b) and C is a convex but subnormal set (see figure 6.4c). Thus B and C are fuzzy sets but not fuzzy numbers.

Zimmermann (1991, p. 85) interpreted a fuzzy number to be normal if it has only one element with membership function equal 1. Consequently he considered a fuzzy set of $\{(3, .8), (4, 1), (5, 1), (6, .7)\}$ as a non-fuzzy number because both $\mu(4) = \mu(5) = 1$. This is not a precise interpretation

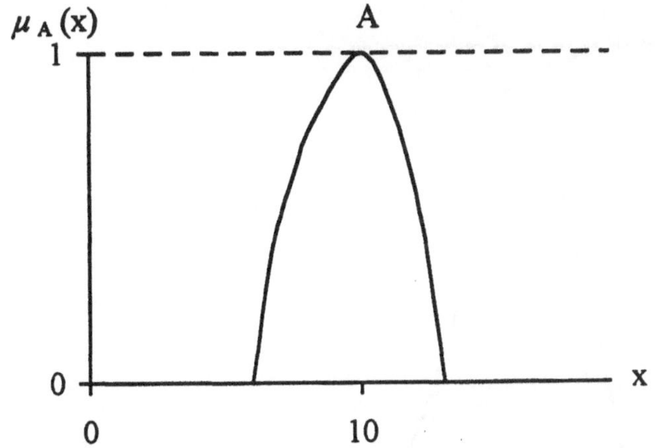

Figure 6.4a Fuzzy number 'approximately 10' (normal and convex)

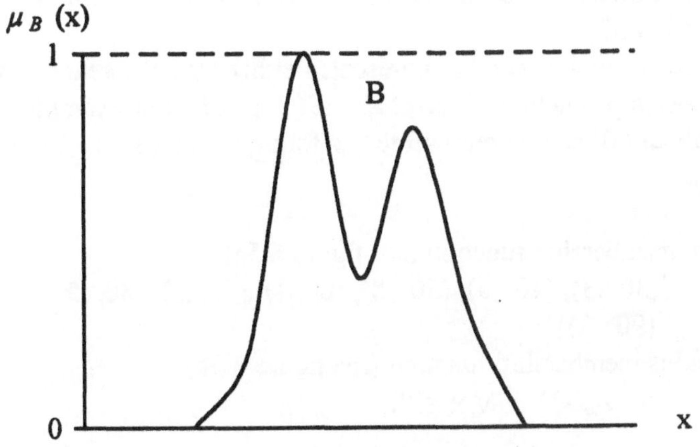

Figure 6.4b Normal but non-convex

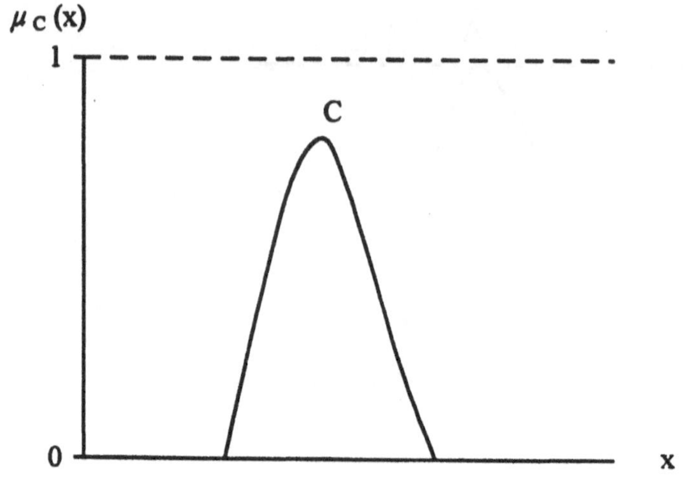

Figure 6.4c Convex but sub-normal

because normality means, as defined earlier, that there is *at least* one element that has membership function equal 1. Moreover, trapezoidal fuzzy numbers (described later) will certainly have more than one element with a membership function of 1.

Further, a fuzzy number can be represented either by a discrete or by a continuous membership function (Chen et al., 1992, p. 67). For example, the fuzzy number 'about 60' can be represented as follows (Chen et al., 1992, p. 69):

- As a discrete membership function (see figure 6.5a):
  'About 60' = {(30, .3), (40, .5), (50, .8), (60, 1), (70, .8), (80, .5), (90, .3)}
- As a continuous membership function (see figure 6.5b):
  'About 60' = $\{x, \mu_{60}(x)\} \quad \forall x \in \Re$

where:

$$\mu_{60}(x) = [1 + ((x - 60) / 10)^2]^{-1}$$

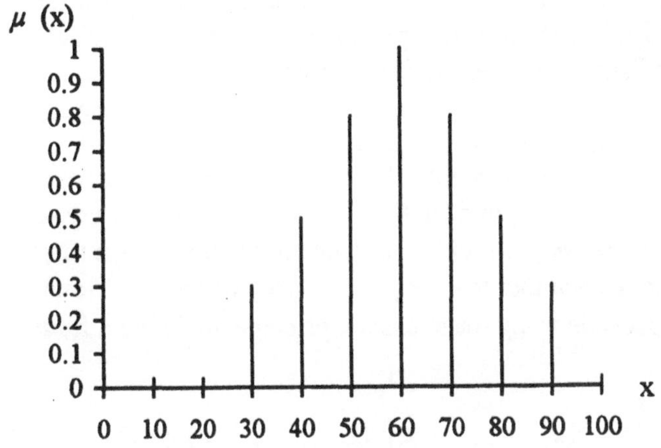

Figure 6.5a A discrete fuzzy number

Source: Chen et al., 1992, p. 68.

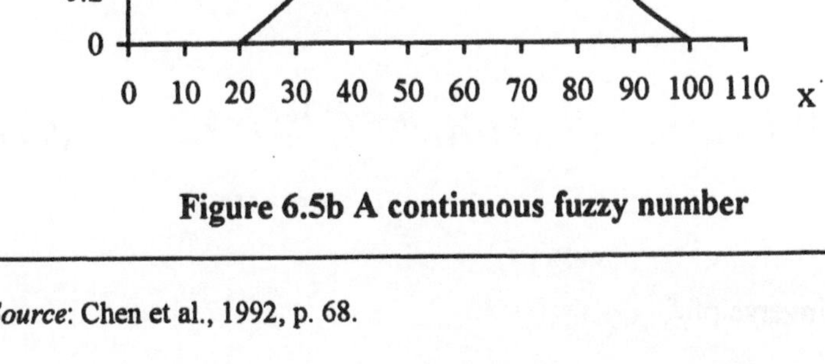

Figure 6.5b A continuous fuzzy number

Source: Chen et al., 1992, p. 68.

The Extension Principle

The extension principle is one of the most basic ideas in fuzzy set theory. Zadeh (1973 and 1975) introduced this principle to extend (generalise) non-fuzzy operations to fuzzy sets. The extension principle can be systematically applied to real algebra, operations on fuzzy numbers, and in defining higher-order fuzzy sets, (Dubois and Prade, 1980, p. 36).

To define the extension principle, let X be a Cartesian product[8] of universes $U = U_1, \ldots\ldots\ldots U_r$, and let $A_1, \ldots\ldots\ldots\ldots A_r$ be r fuzzy sets in $U_1, \ldots\ldots\ldots U_r$, respectively. In addition, define a function f as a mapping from U to a universe Y such that $y = f(x_1, \ldots\ldots, x_r)$ where $y \in Y$ and $x_i \in U_i$, $\forall$ i. Then, it is possible using the extension principle to define a fuzzy set B in Y as follows[9]:

$$B = \{(y, \mu_B(y)) \mid y = f(x_1, \ldots\ldots\ldots x_r), (x_1, \ldots\ldots\ldots x_r) \in U\} \qquad (6.26)$$

where:

$$\mu_B(y) = \begin{cases} \sup\limits_{y = f(x_1,\ldots,x_r)} \min\{\mu_{A_1}(x_1),\ldots\ldots\mu_{A_r}(x_r)\} & if \quad f^{-1}(y) \neq 0 \\ 0 & otherwise \end{cases} \qquad (6.27)$$

where f^{-1} is the inverse of f.

For r = 1, the extension principle could be reduced to

$$B = f(A) = \{(y, \mu_B(y) \mid y = f(x), x \in X\} \qquad (6.28)$$

where:

$$\mu_B(y) = \begin{cases} \min\{\mu_{A_1}(x_1),\ldots\ldots\mu_{A_r}(x_r)\} & if \quad f^{-1}(y) \neq 0 \\ 0 & otherwise \end{cases} \qquad (6.29)$$

where f^{-1} is the inverse of f.

As an example of how the extension principle can extend a crisp algebraic operation into a fuzzy one, let A_1 and A_2 be fuzzy sets as follows (Chen et al., 1992, p. 65):

$$A_1 = \{(3, .4), (4, 1), (5, .7)\}$$
$$A_2 = \{(3, .1), (4, .8), (5, 1), (6, .3)\}$$

and

$$f(x) = 2x_1 + x_2$$

Using the extension principle, $B = f(x) = 2x_1 + x_2$ can be determined as follows:

$$B = \{(9, .1), (10, .4), (11, .4), (12, .8), (13, 1), (14, .7), (15, .7), (16, .3)\}$$

For instance, to calculate $\mu_B(11)$, the possible (x_1, x_2) pairs such that $11 = 2x_1 + x_2$ are (3, 5) and (4, 3). The corresponding membership functions are (.4, 1), (1, .1). The minimum membership in the first pair is .4 and in the second pair is .1. Then $\mu_B(11) = \max(.4, .1) = .4$.

Algebraic Operations on Fuzzy Numbers

The normal algebraic operations such as addition, subtraction, multiplication, and division can be extended to fuzzy numbers using the extension principle. In general, the extension can be achieved by using an interval of confidence (or α-cut level).

The interval of confidence (or α-cut level) can be described by considering a fuzzy number A as illustrated in figure (6.6), A = {(3, 0), (4, .5), (5, .8), (6, 1), (7, .5), (8, 0)}. Every level of α represents a range of values depending on the membership grade or 'degree of belonging' specified. For example, if one presumes the highest level '1', the number is 6, but , on the other hand, if the lowest level '0' is presumed, the number is somewhere between 3 and 8. Kaufmann and Gupta (1985) called these ranges 'intervals of confidence' as they reflect the amount of confidence towards a specific number. Usually, the interval of confidence is written as [a, b] where 'a' and 'b' refer to the left (or minimum) and the right (or maximum) value of the number. For example:

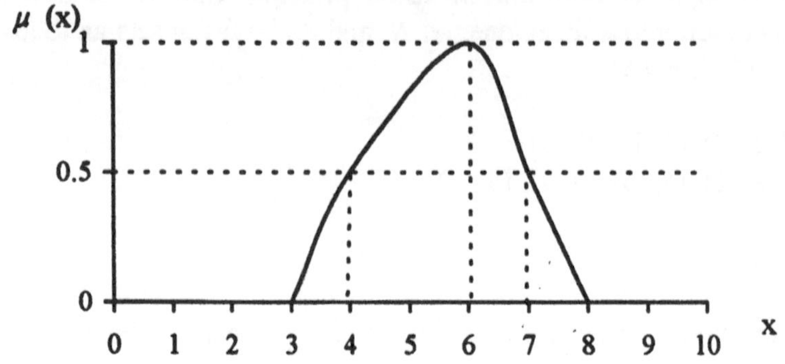

Figure 6.6 A fuzzy number 'approximately 6'

$\alpha_0 = [3, 8]$
$\alpha_{.5} = [4, 7]$
$\alpha_1 = [6, 6]$

Following Kaufmann and Gupta, 1985 and 1991; Zimmermann, 1991; Dubois and Prade, 1980; and Chen et al., 1992, algebraic operations can be extended to fuzzy numbers as follows[10]:

Addition of Fuzzy Numbers Let A, B be fuzzy numbers in $\Re$ defined by their α- cut as follows:

$$A_\alpha = [a_1, a_2] \tag{6.30}$$

$$B_\alpha = [b_1, b_2] \tag{6.31}$$

Then,

$$A_\alpha (+) B_\alpha = [a_1 + b_1, a_2 + b_2] \tag{6.32}$$

As a numerical example, let 'A' represent 'real numbers close to 5' and 'B' represent 'real numbers close to 3' (see figure 6.7), where:

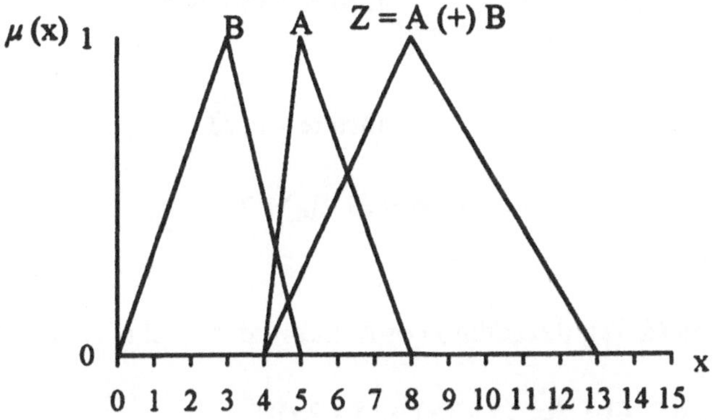

Figure 6.7 Fuzzy numbers A, B, and A (+) B

$$\mu_A(x) = \begin{cases} 0 & x \le 4 \\ (x-4) & 4 < x \le 5 \\ (8-x)/3 & 5 \le x \le 8 \\ 0 & x \ge 8 \end{cases}$$

$$\mu_B(y) = \begin{cases} 0 & y \le 0 \\ y/3 & 0 < y \le 3 \\ (5-y)/2 & 3 \le y \le 5 \\ 0 & y \ge 5 \end{cases}$$

The addition of A and B can be carried out as follows:
The α- cuts of A and B are:

$A_\alpha = [a_1, a_2]$, $B_\alpha = [b_1, b_2]$

Using the membership function at some α level to obtain the interval of confidence as follows:

For A:

if α is set as x - 4, then $\alpha = a_1 - 4$

$\therefore a_1 = \alpha + 4$

if α is set as (8 - x) / 3, then $\alpha = (8 - a_2) / 3$

$\therefore a_2 = 8 - 3\alpha$

For B:

if α is set as y /3, then $\alpha = b_1 /3$

$\therefore b_1 = 3\alpha$

if α is set as (5 - y) / 2, then $\alpha = (5 - b_2) / 2$

$\therefore b_2 = 5- 2\alpha$

Using the equation (6.32), the addition of A and B at α level is computed as:

$$A (+) B = [(\alpha + 4) + 3\alpha, (8 - 3\alpha) + (5 - 2\alpha)]$$
$$= [4\alpha + 4, 13 - 5\alpha]$$

Let

$$Z_\alpha = [z_1, z_2] = [4\alpha + 4, 13 - 5\alpha],$$

then

$$\alpha = (z_1 - 4)/ 4 \text{ and } \alpha = (13 - z_2) / 5$$

Consequently, the addition of A (+) B is

$$Z = \{(z, \mu_\alpha(z)\}$$

where (see figure 6.7):

$$\mu_Z(z) = \begin{cases} 0 & z \le 4 \\ (z-4)/4 & 4 < z \le 8 \\ (13-z)/5 & 8 \le z \le 13 \\ 0 & z \ge 13 \end{cases}$$

The addition of fuzzy numbers has the following properties (Chen et al., 1992, p. 70):

1. Commutative:

$$A(+)\,B = B(+)\,A$$

2. Associative:

$$[A\,(+)\,B]\,(+)\,C = A\,(+)\,[B\,(+)\,C]$$

3. If a neutral exists at the left and the right, it is the real number 0. Thus:

$$A\,(+)\,0 = 0\,(+)\,A = A.$$

4. Nonsymmetric:

$$A\,(+)\,(-B) = (-B)\,(+)\,A \neq 0$$

where -B is the image of B with membership function:

$$\mu_{-B}(x) = \mu_B(x)$$

Subtraction of Fuzzy Numbers The subtraction of fuzzy numbers A and B is defined as follows:

$$A_\alpha\,(-)\,B_\alpha = [\,a_1 - b_2,\, a_2 - b_1] \tag{6.33}$$

Using the same numerical example as in the addition case, the subtraction of B from A at α level is computed as:

$$\begin{aligned} A\,(-)\,B &= [\,(\alpha + 4) - (\,5 - 2\alpha),\,(8 - 3\alpha) - 3\alpha] \\ &= [\,3\alpha - 1,\, 8 - 6\alpha] \end{aligned}$$

Let:

$$Z_\alpha = [z_1, z_2] = [\,3\alpha - 1,\, 8 - 6\alpha]$$

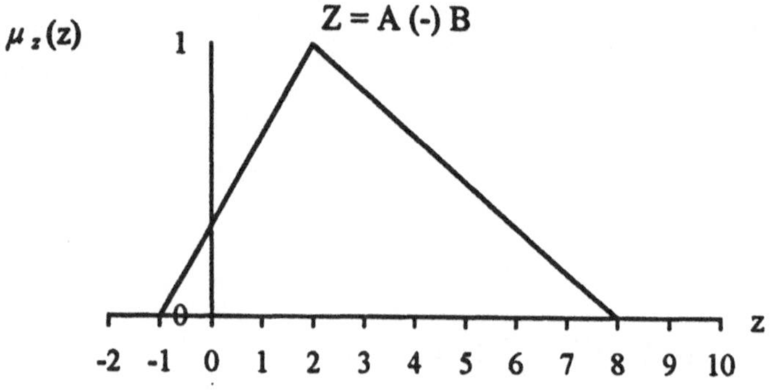

Figure 6.8 Subtraction of two fuzzy numbers A and B

Then,

$\alpha = (z_1 + 1) / 3$ and $\alpha = (8 - z_2) / 6$

Consequently, the subtraction of A (–) B is:

$Z = \{(z, \mu_z(z)\}$

where (see figure 6.8):

$$\mu_Z(z) = \begin{cases} 0 & z \leq 1 \\ (z+1)/3 & -1 < z \leq 2 \\ (8-z)/6 & 2 \leq z \leq 8 \\ 0 & z \geq 8 \end{cases}$$

The commutative and associative properties are not applicable in the case of subtraction (Kaufmann and Gupta, 1985, p. 22).

Multiplication of Fuzzy Numbers The multiplication of the positive fuzzy numbers A and B is defined as:

$$A_\alpha\,(\bullet)\,B_\alpha = [\,a_1\,b_1,\ a_2\,b_2] \tag{6.34}$$

Using the same numerical example as before, the product of A and B at α level is computed as:

$$\begin{aligned} A\,(\bullet)\,B &= [(\alpha + 4)\,(3\alpha),\ (8 - 3\alpha)\,(5 - 2\alpha)] \\ &= [\,3\alpha^2 + 12\alpha,\ 6\alpha^2 - 31\alpha + 40] \end{aligned}$$

Let:

$$Z_\alpha = [z_1,\ z_2] = [\,3\alpha^2 + 12\alpha,\ 6\alpha^2 - 31\alpha + 40]$$

Then:

$$3\alpha^2 + 12\alpha - z_1 = 0$$

and:

$$6\alpha^2 - 31\alpha + 40 - z_2 = 0$$

The roots of these two equations are, respectively:

$$\alpha = (-12 + (144 + 12z_1)^{.5})\,/\,6$$
$$\alpha = (31 + (1 + 24z_2)^{.5})\,/\,12$$

Consequently, the product of A $(\bullet)$ B is:

$$Z = \{z,\ \mu_Z(z)\}$$

where (see figure 6.9):

$$\mu_Z(z) = \begin{cases} 0 & z \le 0 \\ (-12 + (144 + 12z)^{.5})/6 & 0 < z \le 15 \\ (31 + (1 + 24z)^{.5})/12 & 15 \le z \le 40 \\ 0 & z \ge 40 \end{cases}$$

It should be noted that Z is still a fuzzy number even though its membership function is no longer linear.

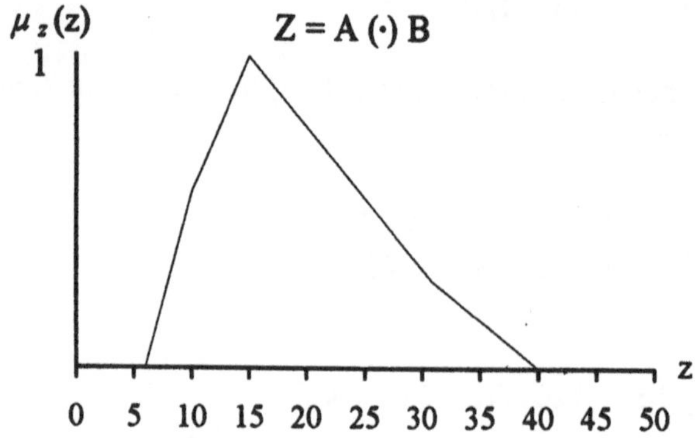

Figure 6.9 Multiplication of two fuzzy numbers A and B

Fuzzy multiplication has the following properties (Chen et al., 1992, p. 77):

1. When both A and B have the same sign, A (•) B is a positive fuzzy number.

2. (-A) (•) B = - (A (•) B)

3. The multiplication of fuzzy numbers A and B is commutative and associative, i.e.,

   A (•) B = B (•) A

and:

   (A (•) B) (•) C = A (•) (B (•) C)

4. If a neutral exists at the left and at the right, it is the real number 1, i.e.,

   A (•) 1 = 1 (•) A = A

5. The inverse of A is A^{-1} and A (•) $A^{-1} \neq 1$

where:

$$A_\alpha^{-1} = [\ 1/a_1,\ 1/a_2] \tag{6.35}$$

The Division of Fuzzy Numbers The division of positive fuzzy numbers A and B can be defined as:

$$A_\alpha\ (\div)\ B_\alpha = [a_1 / b_2,\ a_2 / b_1] \qquad b_2 > 0 \tag{6.36}$$

The division operation is an extension of multiplication i.e.,

$$A\ (\div)\ B = A\ (\bullet)\ B^{-1} \tag{6.37}$$

where:

$$A\ (\bullet)\ B^{-1} = [a_1\ (1/b_2),\ a_2\ (1/b_1)] \tag{6.38}$$

Multiplication of a Fuzzy Number by an Ordinary Number Let A be a fuzzy number in $\Re$ and k a positive ordinary (crisp) number, then:

$$\begin{aligned} k\ (\bullet)\ A_\alpha &= [k, k]\ (\bullet)\ [a_1, a_2] \\ &= [ka_1, ka_2] \end{aligned} \tag{6.39}$$

In general, it will be noted that the distributive property is preserved for fuzzy operations i.e., given three positive fuzzy numbers A, B, and C, then:

(A (+) B) (•) C = (A (•) C) + (B (•) C) (Dubois and Prade, 1980)

Special Fuzzy Numbers

The implementation of algebraic operations involving fuzzy numbers is thus likely to involve extensive computations. However, for practical purposes, some restrictions may be placed on the type of the membership function. Such restrictions result in special kinds of fuzzy numbers: L-R fuzzy numbers, triangular fuzzy numbers, and trapezoidal fuzzy numbers, and these special numbers allow considerable simplification of computing effort.

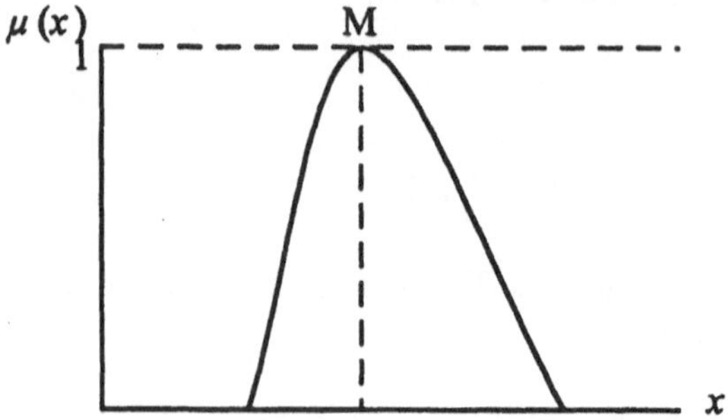

Figure 6.10 L-R fuzzy number

L-R Fuzzy Numbers Dubois and Prade (1980, pp 53-54) define a fuzzy number M to be of type L-R if it has the membership function:

$$\mu_M(x) = \begin{cases} L((m-x)/\alpha) & \forall\, x \le m, \quad \alpha > 0 \\ R((x-m)/\beta) & \forall\, x \ge m, \quad \beta > 0 \end{cases} \tag{6.40}$$

Where L and R are reference functions for the left and the right respectively, m is the mean value[11] of M, and α and β are called the left and right spreads, respectively (see figure 6.10). When $\alpha = \beta =$ 0, M is a non-fuzzy number. As the spreads increase, M becomes fuzzier. Symbolically, the L-R fuzzy number can be written:

$$\mathrm{M} = (m, \alpha, \beta) \tag{6.41}$$

It should be noted that determining the reference functions L and R depends on the nature of the phenomenon to which the fuzzy number refers. For example let:

$L(x) = 1 / (1+x^2)$
$R(x) = 1 / (1 + 2\,|x|)$
$\alpha = 2, \beta = 3, m = 5$

Then:

$$\mu_5(x) = \begin{cases} L\left(\frac{5-x}{2}\right) = \dfrac{1}{1+\left(\frac{5-x}{2}\right)^2} & \forall x \le 5, \\ R\left(\frac{x-5}{3}\right) = \dfrac{1}{1+\left|\frac{2(x-5)}{3}\right|} & \forall x \ge 5 \end{cases}$$

Algebraic operations on L-R fuzzy numbers can be carried out as follows (Chen et al., 1992, p. 90).

Given two L-R fuzzy numbers defined as:

M = (m, α, β)
N = (n, ε, δ)

and an ordinary number k.

where m and n are the means values, α and ε are the left hand spreads and β and δ are the right hand spreads of M and N, respectively.

1. Image of N:

$$-N = (-n, \delta, \varepsilon) \tag{6.42}$$

2. Inverse of N:

$$N^{-1} = (n^{-1}, \delta/(n(n+\delta)), \varepsilon/(n(n-\varepsilon))) \qquad n \neq 0 \tag{6.43}$$

3. Addition:

$$M\,(+)\,N = (m+n, \alpha+\varepsilon, \beta+\delta) \tag{6.44}$$

4. Subtraction:

$$M\,(-)\,N = (m-n, \alpha+\delta, \beta+\varepsilon) \tag{6.45}$$

5. Scalar multiplication:

$$k > 0, k \in \Re: \quad k\,(\bullet)\,M = (km, k\alpha, k\beta) \tag{6.46}$$
$$k < 0, k \in \Re: \quad k\,(\bullet)\,M = (km, -k\beta, -k\alpha) \tag{6.47}$$

Table 6.2 An example for algebraic operations on L-R fuzzy numbers

M	(5, 1, 2)
N	(8, 3, 2)
M (+) N	(13, 4, 4)
M (−) N	(-3, 3, 5)
M (•) N	(40, 20, 22)
M (÷) N	(5/8, 9/40, 31/40)

6. Multiplication:

$$M \geq 0, N \geq 0: M\ (\bullet)\ N = (mn,\ m\varepsilon + n\alpha - \alpha\varepsilon,\ m\delta + n\beta - \beta\delta) \qquad (6.48)$$

7. Division:

$$M \geq 0, N > 0: M\ (\div)\ N = (m/n,\ (m\delta + n\alpha)/(n(n+\delta)),\ (m\varepsilon + n\beta)/(n(n - \varepsilon))) \qquad (6.49)$$

Table (6.2) gives a numerical example to illustrate basic algebraic operations on L-R fuzzy numbers.

Dubois and Prade (1980, p. 57) demonstrated a case in which '*m*' is an interval [*m*, *p*] rather than a unique number. This case is called an L-R flat fuzzy number. They defined the membership function of this type as follows:

$$\mu_M(x) = \begin{cases} L((m-x)/\alpha) & \forall\ x \leq m,\ \alpha > 0 \\ R(x-p)/\beta) & \forall\ x \geq p,\ \beta > 0 \\ 1 & m \leq x \leq p \end{cases} \qquad (6.50)$$

The L-R flat fuzzy number can be written as follows (see figure 6.11):

$$M = (m, p, \alpha, \beta) \qquad (6.51)$$

As noted by Zimmermann (1990, p. 64) this definition is very general and allows the quantification of quite different types of information. For instance, if M is a real crisp number, $m \in \Re$,

$$M = (m, m, 0, 0) \quad \forall L, \forall R$$

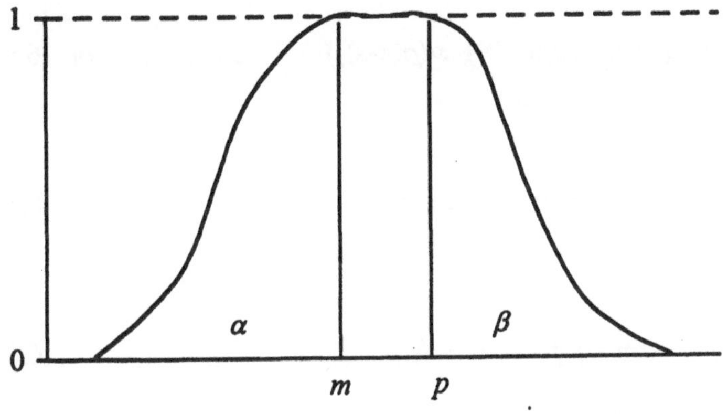

Figure 6.11 An L-R flat fuzzy number

If M is a crisp interval $[m, p]$

$$M = (m, p, 0, 0) \qquad \forall L, \forall R$$

Algebraic operations of L-R flat fuzzy numbers are as follows (Chen et al., 1992, p. 91):

Given two L-R flat fuzzy numbers defined as:

$$M = (m,\ p,\ \alpha,\ \beta)$$
$$N = (n,\ q,\ \varepsilon,\ \delta)$$

and an ordinary number k.

1. Image of N:

$$-N = (-q,\ -n,\ \delta,\ \varepsilon) \tag{6.52}$$

2. Inverse of N:

$$N^{-1} = (1/q,\ 1/n,\ \delta/(q(q+\delta)),\ \varepsilon/(n(n-\varepsilon))) \qquad 0 \notin (n,\ q,\ \varepsilon,\ \delta) \quad (6.53)$$

3. Addition:

$$M\ (+)\ N = (m + n,\ p + q,\ \alpha + \varepsilon,\ \beta + \delta) \quad (6.54)$$

4. Subtraction:

$$M\ (-)\ N = (m - q,\ p - n,\ \alpha + \delta,\ \beta + \varepsilon) \quad (6.55)$$

5. Multiplication:

$$M \geq 0, N \geq 0\text{: } M\ (\bullet)\ N = (mn,\ pq,\ m\varepsilon + n\alpha - \alpha\varepsilon,\ p\delta + q\beta + \beta\delta) \quad (6.56)$$

6. Division:

$$M \geq 0, N > 0\text{: } M\ (\div)\ N = (mq,\ p/n,\ (m\delta+q\alpha)/(q(q+\delta)),\ (p\varepsilon+n\beta)/(n(n-\varepsilon))) \quad (6.57)$$

Table 6.3 gives a numerical example to illustrate basic algebraic operations on L-R flat fuzzy numbers.

It might be noted that L-R flat fuzzy numbers M = (m, α, β) are actually a special case of L-R flat fuzzy numbers M = (m, m, α, β). The fuzzy number M = (m, α, β) might be interpreted as approximately m, while M = (m_1, m_2, α, β) could be interpreted as approximately between m_1 and m_2.

Table 6.3 A numerical example for L-R flat fuzzy numbers

M	(3, 5, 2, 3)
N	(2, 4, 1, 2)
M (+) N	(5, 9, 3, 5)
M (–) N	(-1, 3, 0, 2)
M (•) N	(6, 20, 5, 16)
M (÷) N	(3/4, 5/2, 7/12, 11/2)

Triangular Fuzzy Numbers Laarhoven and Pedrycz (1983) introduced another special kind of fuzzy number: the triangular fuzzy number. They define a fuzzy number M in $\Re$ to be a triangular number if its membership function μ_M: $\Re \rightarrow [0,1]$ is equal to:

$$\mu_M(x) = \begin{cases} 0 & x \leq a \\ (x-a)/(b-a) & a \leq x \leq b \\ (c-x)/(c-b) & b \leq x \leq c \\ 0 & x \geq c \end{cases} \tag{6.58}$$

where a, b, c $\in \Re$ and a $\leq$ b $\leq$ c (see figure 6.12).

A triangular fuzzy number can be defined as follows:

$$M = (a, b, c) \tag{6.59}$$

Operations on triangular fuzzy numbers can be carried out as follows (Laarhoven and Pedrycz, 1983; Kaufmann and Gupta, 1990; and Chen et al.1992).

Given two triangular fuzzy numbers M and N defined as:

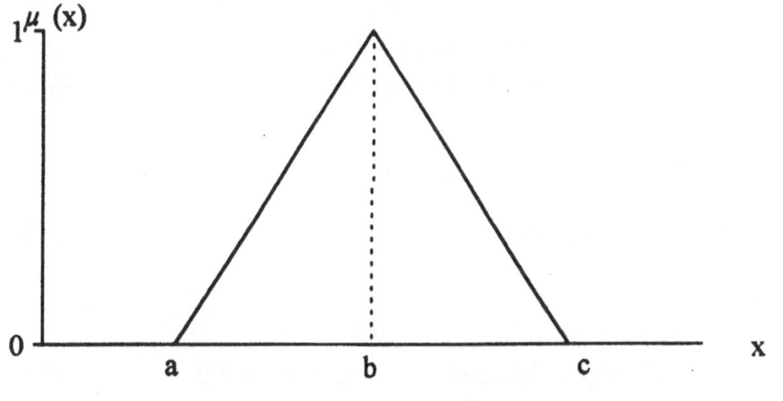

Figure 6.12 A triangular fuzzy number

$M = (a, b, c)$
$N = (e, f, g)$

and an ordinary number k.

1. Image of N:

$$-N = (-g, -f, -e) \tag{6.60}$$

2. Inverse of N:

$$N^{-1} = (1/g, 1/f, 1/e) \qquad 0 \notin [e, g] \tag{6.61}$$

3. Addition:

$$M (+) N = (a + e, b + f, c + g) \tag{6.62}$$

4. Subtraction:

$$M (-) N = (a -g, b- f, c- e) \tag{6.63}$$

5. Scalar Multiplication:

$$\forall k > 0, k \in \Re: k (\bullet) M = (ka, kb, kc) \tag{6.64}$$
$$\forall k < 0, k \in \Re: k (\bullet) M = (kc, kb, ka) \tag{6.65}$$

6. Multiplication:

$$M (\bullet) N = (ae, bf, cg) \qquad a \geq 0, \quad e \geq 0 \tag{6.66}$$

7. Division:

$$M (\div) N = (a/g, b/f, c/e) \qquad a \geq 0, \quad e > 0 \tag{6.67}$$

Table (6.4) gives a numerical example to illustrate algebraic operations on triangular fuzzy numbers.

Table 6.4 An example for algebraic operations on triangular fuzzy numbers

M	**(10, 12, 15)**
N	**(8, 11, 14)**
M (+) N	(18, 23, 29)
M (−) N	(-4, 1, 7)
M (•) N	(80, 132, 210)
M (÷) N	(5/7, 12/11, 15/8)

Trapezoidal Fuzzy Numbers A trapezoidal fuzzy number M can be defined as a quadruplet (a, b, c, d) which has the following membership function, see figure 6.13, (Kaufmann and Gupta, 1991, p. 32).

$$\mu_M(x) = \begin{cases} 0 & x \le a \\ (x-a)/(b-a) & a \le x \le b \\ 1 & a \le x \le c \\ (d-x)/(d-c) & c \le x \le d \\ 0 & x \ge d \end{cases} \tag{6.68}$$

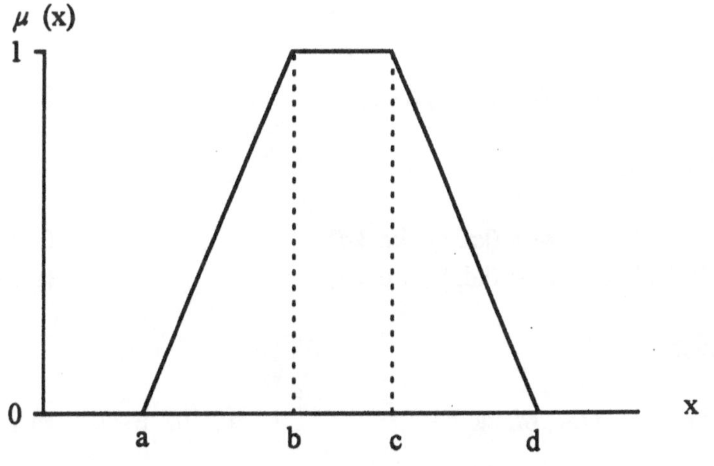

Figure 6.13 A trapeziodal fuzzy number

The algebraic operations on trapezoidal fuzzy numbers are as follows (Chen et al., 1992, p. 93).

Given two trapezoidal fuzzy numbers M and N defined as:

$M = (a, b, c, d)$
$N = (e, f, g, h)$

and an ordinary number k.

1. Image of N:

$$-N = (-h, -g, -f, -e) \tag{6.69}$$

2. Inverse of N:

$$N^{-1} = (1/h,\ 1/g,\ 1/f,\ 1/e) \qquad 0 \notin [e, h] \tag{6.70}$$

3. Addition:

$$M\ (+)\ N = (a + e, b + f, c + g, d + h) \tag{6.71}$$

4. Subtraction:

$$M\ (-)\ N = (a - h, b - g, c - f, d - e) \tag{6.72}$$

5. Scalar multiplication:

$$\forall\, k > 0, k \in \Re: k\ (\bullet)\ M = (ka, kb, kc, kd) \tag{6.73}$$
$$\forall\, k < 0, k \in \Re: k\ (\bullet)\ M = (kd, kc, kb, ka) \tag{6.74}$$

6. Multiplication:

$$M\ (\bullet)\ N = (ae, bf, cg, dh) \qquad a \geq 0,\ \ e \geq 0 \tag{6.75}$$

Table 6.5 A numerical example for trapezoidal fuzzy numbers

M	**(12, 15, 18, 20)**
N	**(9, 10, 12, 14)**
M (+) N	(21, 25, 30, 34)
M (−) N	(-2, 3, 8, 11)
M (•) N	(108, 150, 216, 280)
M (÷) N	(6/7, 5/4, 9/5, 20/9)

7. Division:

$$M\ (\div)\ N = (a/h,\ b/g,\ c/f,\ d/e) \qquad a \geq 0,\ \ e > 0 \qquad (6.76)$$

Table 6.5 gives a numerical example to illustrate algebraic operations on trapezoidal fuzzy numbers.

It should be noted that a triangular fuzzy number is a special case of a trapezoidal fuzzy number with b = c. Also, a trapezoidal fuzzy number can be interpreted as a fuzzy quantity 'approximately between b and c'.

It can be seen that both triangular and trapezoidal numbers are special cases of L-R fuzzy numbers in which the L and R functions are linear.

Ranking Fuzzy Numbers

The issue of ranking or ordering fuzzy numbers is very important in the context of decision analysis. As fuzzy numbers represent imprecise numerical quantities, it can be difficult to ascertain whether one fuzzy number is 'better' or 'worse', than another. For example, in figure 6.14 three fuzzy numbers A, B and C represent a final rating for three alternative choices; the question is which one is the best? Alternative methods of tackling this question have been proposed in the literature of fuzzy set theory. These methods can be classified into three approaches: the associated ordinary number approach, the maximising set approach, and the maximising and minimising sets approach. These approaches will first be reviewed, and then a new ranking method that uses more of the information contained in fuzzy numbers, will be proposed.

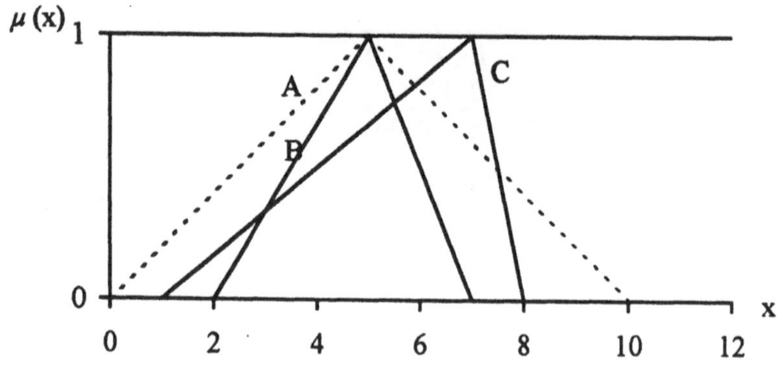

Figure 6.14 Three alternative choices represented by their fuzzy numbers

The Associated Ordinary Number Approach

A fuzzy number can be represented by an associated ordinary number that, for a triangular fuzzy number A = (a, b, c), could be defined as a weighted average as follows (Kaufmann and Gupta, 1991, p. 38):

$$A^{\backslash} = (a + 2b + c)/4 \qquad (6.77)$$

Kaufmann and Gupta (1991, pp. 37- 41) suggested using the associated ordinary number as a first criterion to rank fuzzy numbers. If two or more fuzzy numbers have the same associated ordinary number, then a second criterion can be used: the mode (b) of each of the fuzzy numbers. If the mode fails to distinguish between two or more fuzzy numbers, then a third criterion is used: the divergence. The divergence of a fuzzy number is calculated as the difference between 'a' and 'c', i.e. the fuzzy numbers are ranked using (c-a); smaller divergences being ranked higher than lager ones.

To clarify the criteria in Kaufmann and Gupta's approach, suppose for example, there are four fuzzy numbers:

A = (0, 6, 7)
B = (-3, 5, 11)

C = (-2, 6, 8)
D = (-1, 6, 7)

The associated ordinary numbers (calculated using 6.92):

A = (0 +2(6) +7)/4 = 4.75
B= (-3 + 2(5) + 11)/4 = 4.5
C = (-2 +2(6) + 8) /4 = 4.5
D = (-1 +2(6) + 7) /4= 4.5

According to this criterion, the order of these numbers is first A and then B, C, and D, which are indistinguishable.

The second criterion is therefore needed in order to distinguish between B, C and D.

The mode

B = (-3, 5, 11):	mode = 5
C = (-2, 6, 8):	mode = 6
D = (-1, 6, 7):	mode = 6

According to the first two criteria, the order is A, then C and D and finally then B.

The third criterion is therefore needed to distinguish between C and D.

The divergence

C = (-2, 6, 8):	divergence = 8 + 2 = 10
D = (-1, 6, 7):	divergence = 7 + 1 = 8

Thus the order of the four fuzzy numbers is A > D > C > B

It should be noted that the three criteria suggested by Kaufmann and Gupta must be followed in sequence. If the three criteria fail to distinguish between fuzzy numbers, then they are equal.

Recently, Chiu and Park (1994) have suggested a weighted method of ranking rather similar to the first criterion used in Kaufmann and Gupta's method. They used the following formula, again using the triangular fuzzy number A = (a, b, c):

$$[(a + b + c) / 3] + wb \tag{6.78}$$

where w is the relative weight determined by the nature and the magnitude of the most promising value (b). They suggest that if the magnitude of the most promising value (b) is important, a larger number for w, such as .3, is recommended, otherwise, a smaller weight, such as 0.1, is recommended.

Using Chui and Park's method, the four fuzzy numbers in the previous example can be ranked as follows (w = 0.5 for the purpose of illustration):

A = (0, 6, 7) = (0 + 6 + 7) /3 + .5(6) = 7.3
B = (-3, 5, 11) = (-3 +5 + 11) /3 + .5(5) = 6.8
C = (-2, 6, 8) = (-2 + 6 + 8) /3 + .5(6) = 7
D = (-1, 6, 7) = (-1 + 6 + 7) /3 + .5(6) = 7

∴ The order is A > C = D > B

As is clear from this example, the method does not distinguish between C and D, as Kaufmann and Gupta's does after applying their third criterion (the divergence). This is because Chui and Park's method combines the first two criteria in Kaufmann and Gupta's method in a slightly different way but ignores the third criterion. However, if w = 0.1, the ranking is A>B>C=D.

The Maximising Set Approach

Jain (1976 and 1977) developed a ranking method using the concept of a maximising set. He defined the maximising set M(y) of a set Y as:

> '... a fuzzy set such that the grade of membership of a point y ∈ Y in M(Y) represents the degree to which y approximates to sup Y in some specified sense'. (Jain, 1976, p. 700)

In order to use the maximising set approach, first a set Y is formed that is a union of the supports for each of the alternatives (A_i). Formally:

$$Y = \bigcup_{i=1}^{n} S(A_i) \qquad i = 1, \ldots\ldots n \tag{6.79}$$

where $S(A_i)$ is the support of fuzzy numbers associated with alternative A_i.

Then, the maximising set $M(A_i)$ for the alternative A_i is obtained, which is:

$$M(A_i) = \{ x, \mu_M(A_i) \} \qquad (6.80)$$

where:

$$\mu_M(A_i) = [x / x_{max}]^k \qquad (6.81)$$

and:

$$x_{max} = \sup Y \qquad (6.82)$$

and k is an integer and depends on the application[12].

After determining the maximising set, an optimal fuzzy set is formed for each alternative (fuzzy number), A_i as $O(A_i)$ which represents the fuzzy intersection of $M(A_i)$ and A_i, hence its grade of membership is characterised as:

$$\mu_{O(Ai)} = \mu_M(A_i) \wedge \mu_x(A_i) \qquad (6.83)$$

The set $O(A_i)$ takes into account the grade of membership of each element in A_i and the relative value of each element at the same time. Thus, the set $O(A_i)$ might represent the grade of membership of each element resulting when alternative A_i is selected. Then the set of optimal alternatives L is determined as follows:

$$L = \{ A_i , \mu_L(A_i)\} \qquad (6.84)$$

where:

$$\mu_L(A_i) = \vee O(A_i) \qquad (6.85)$$

The example below illustrates Jain's approach. Let A_1, A_2 and A_3 be three alternatives represented by fuzzy sets as follows:

$A_1 = \{(1, .7), (2, .88), (3,1), (7, .3)\}$,

$A_2 = \{(1, 1), (6, .3), (7, .82), (8, .8)\}$,
$A_3 = \{(3, .4), (4, .8), (5, .1), (6, .7), (8, .3)\}$

The set Y is determined as follows:

$$Y = S(A_1) \cup S(A_2) \cup S(A_3)$$
$$= \{1, 2, 3, 7\} \cup \{1, 6, 7, 8\} \cup \{3, 4, 5, 6, 8\}$$
$$= \{1, 2, 3, 4, 5, 6, 7, 8\}$$

$\therefore$ sup Y = 8

Assuming k = 1, then the maximising sets for various alternatives are determined as follows:

$$M_1 = \{(1, .125), (2, .25), (3, .375), (7, .875)\}$$
$$M_2 = \{(1, .125), (6, .75), (7, .875), (8, 1)\}$$
$$M_3 = \{(3, .375), (4, .5), (5, .625), (6, .75), (8, 1)\}$$

Then, the optimal fuzzy set for each alternative $O(A_i)$ can be determined as follows:

$$O(A_1) = \{(1, .125), (2, .25), (3, .375), (7, .3)\}$$
$$O(A_2) = \{(1, .125), (6, .3), (7, .82), (8, .8)\}$$
$$O(A_3) = \{(3, .375), (4, .5), (5, .1), (6, .7), (8, .3)\}$$

The optimality degree for each alternative can be determined as follows:

$$\mu_L(A_1) = \vee (.125, .25, .375, .3) = .375$$
$$\mu_L(A_2) = \vee (.125, .3, .82, .8) = .82$$
$$\mu_L(A_3) = \vee (.375, .5, .625, .7, .3) = .7$$

$\therefore$ The set of optimal alternatives is:

$$L = \{(A_1, .375), (A_2, .82), (A_3, .7)\}$$

Thus, the alternatives can be ranked as $A_2 > A_3 > A_1$.

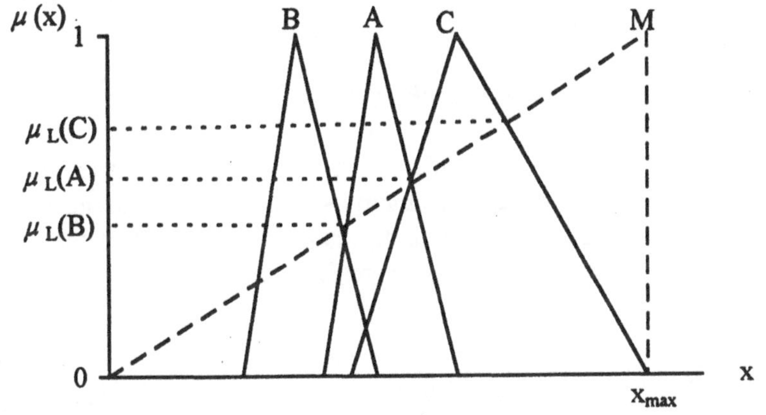

Figure 6.15 The maximising set of Jain's method

Jain's method can be graphically depicted as in figure 6.15, using three triangular fuzzy numbers A, B and C. The ranking of the fuzzy numbers is determined by the intersection of M and the right-hand side of the membership function of each fuzzy number. Accordingly, in figure 6.15 C > A > B.

Jain's method for ranking fuzzy numbers has been criticised by many authors (see for example: Chen, 1985: Kim and Park, 1990; and Chen et al., 1992) for two reasons:

1. The method may give counter-intuitive results. For example in figure 6.16 A and B share the same right hand side membership, accordingly, Jain's method gives the same score to both, that is, A = B, although it is clear that B is better than A, (Chen et al., 1992, p. 237). The reason for this counter-intuitive result is the fact that only part of the information contained in the fuzzy numbers (that is, the non-increasing part or right-hand side of the membership function) is considered. Kim and Park concluded that Jain's method reflects only the optimistic point of view.
2. Chen (1985) noted that Jain's method contradicts the definition of the membership function ($0 \leq \mu(x) \leq 1$) if some of the fuzzy numbers contain negative support. Specifically, he showed that if k is an odd integer, then the membership function of $\mu_x(A_i) = [x/x_{max}]^k$ becomes

negative. Also, if k is an even integer, and if negative support x has $|x| > |x_{max}|$, then $\mu_x(A_i) > 1$.

Minimising and Maximising Approach

To overcome the criticisms of Jain's method, Chen (1985) redefined the maximising set (M) to reflect the minimum as well as the maximum values associated with the fuzzy numbers under consideration. The membership function of M is defined (assuming there are A_i fuzzy numbers to be ranked) as:

$$\mu_M(x) = \begin{cases} [(x - x_{min})/(x_{max} - x_{min})]^k & x_{min} \le x \le x_{max} \\ 0 & otherwise \end{cases} \quad (6.86)$$

where:

$$x_{min} = inf\, S, \quad (6.87)$$

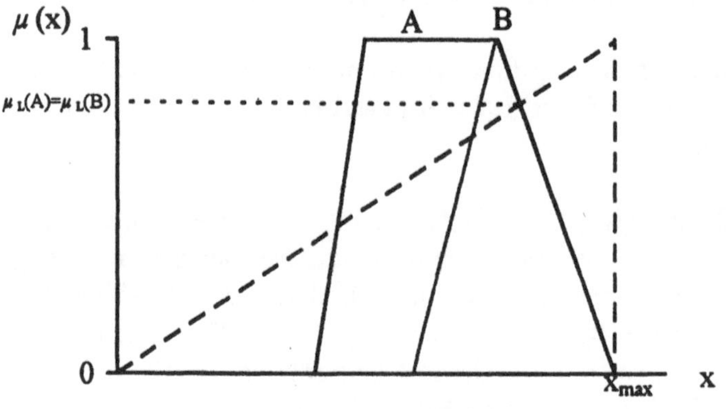

Figure 6.16 A counter-intuitive result using Jain's method

Source: Chen et al., 1992, p. 238.

$$x_{max} = sup\ S, \tag{6.88}$$

$$S = \bigcup_{i=1}^{n} S_i, \tag{6.89}$$

$$S_i = \{x \mid \mu_{Ai}(x) > 0\} \qquad i = 1, \ldots n \tag{6.90}$$

The value of k can be varied to suit the application.

Also, the right score of each fuzzy number A_i is defined:

$$R_M\,(i) = \sup\,(\mu_M(x) \wedge \mu_A(x)) \qquad i = 1,\ldots\ n. \tag{6.91}$$

Thus far Chen's method follows the same procedure as Jain's, however, a minimising set (G) is now introduced to reflect the information in the left-hand side of a fuzzy number in the ranking process. The membership function of 'G' is defined as:

$$\mu_G(x) = \begin{cases} [(x_{\max} - x)/(x_{\max} - x_{\min})]^k & x_{\min} \le x \le x_{\max} \\ 0 & otherwise \end{cases} \tag{6.92}$$

where x_{max} and x_{min} are defined as above.

The left score of each fuzzy number A_i is defined as:

$$L_G\,(i) = \sup\,(\mu_G(x) \wedge \mu_A(x)) \qquad i = 1,\ldots\ n. \tag{6.93}$$

The maximising and minimising sets as defined in Chen's method are illustrated in figure 6.17.

The total score for each fuzzy number is determined as:

$$T(i) = [R_M(i) + (1 - L_G(i))]\,/\,2 \tag{6.94}$$

It should be noted that 1- $L_G(i)$ is used to reverse the effect of the $L_G(i)$ value because a higher $L_G(i)$ value indicates a worse fuzzy number. Division by 2 makes the total score fall in the range [0,1]. Thus, the higher the value

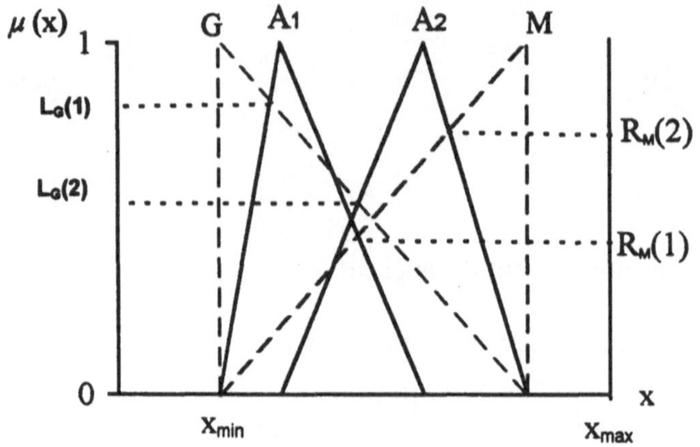

Figure 6.17 The maximising and minimising sets of Chen's method

of T(i), the more preferred is the fuzzy number A_i (Chen et al., 1992, p. 240).

As an example to illustrate Chen's method, suppose that there are three fuzzy numbers $A_1 = (4,5,6)$, $A_2 = (3,6,7)$, and $A_3 = (5,8,10)$ as in figure 6.18.

Assuming k =1, the maximising and minimising sets can be determined as:

$$\mu_M(x) = \begin{cases} (x-3)/7 & 3 \le x \le 10 \\ 0 & otherwise \end{cases}$$

$$\mu_G(x) = \begin{cases} (10-x)/7 & 3 \le x \le 10 \\ 0 & otherwise \end{cases}$$

Table 6.6 gives the $R_M(i)$, $L_G(i)$ and T(i). Accordingly, the ranking is $A_3 > A_2 > A_1$.

Chen et al. (1992, p. 245) pointed out that Chen's method ignores the absolute locations of fuzzy numbers on the x-axis. For instance, assuming there are two fuzzy numbers A_1 and A_2 as in figure 6.19a and A'_1 and A'_2

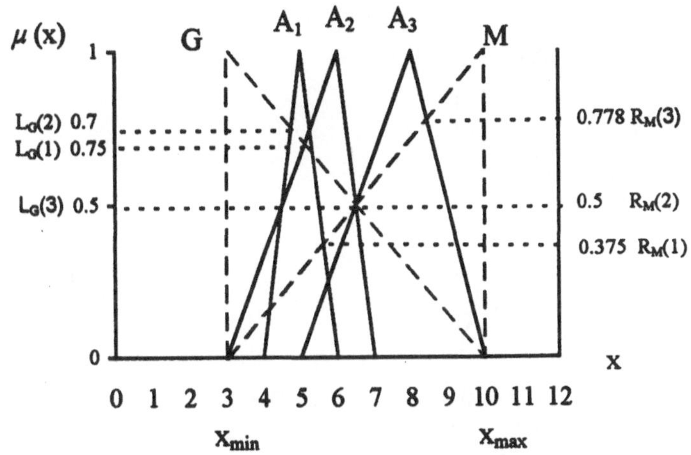

Figure 6.18 An example of Chen's method

Table 6.6 The R_M(i), L_G(i) and T(i) of the three alternatives

	R_M(i)	L_G(i)	T(i)
A_1	0.375	0.750	0.313
A_2	0.500	0.700	0.400
A_3	0.778	0.500	0.679

as in figure 6.19b. According to Chen's method, the total scores of these fuzzy numbers are $T(A_1) = T(A'_1) = .417$ and $T(A_2) = T(A'_2) = .545$. Since A_1 and A'_1 are obviously different fuzzy numbers, their T(i) scores should not be the same. The equality is attributed to the fact that this method considers only the relative location of fuzzy numbers and not their absolute location on the real line $\Re$.

To resolve the above problem, Chen and Hwang (1989)[13] redefined the maximising and minimising sets as in figure 6.20.

The membership functions of the maximising and minimising sets are defined as follows:

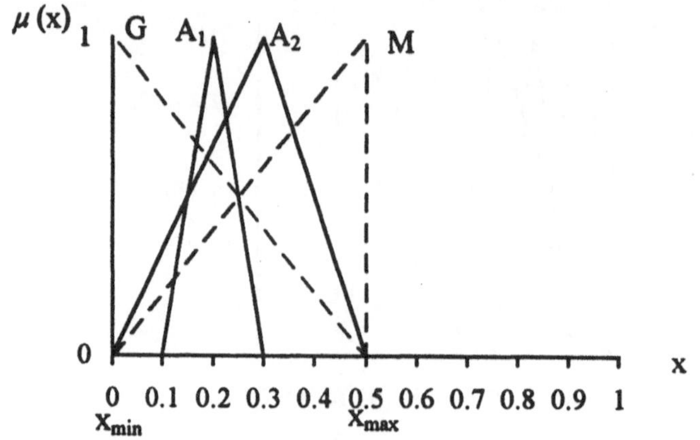

Figure 6.19a An example of the illogicality in Chen's method

Source: Chen et al., 1992, p. 246.

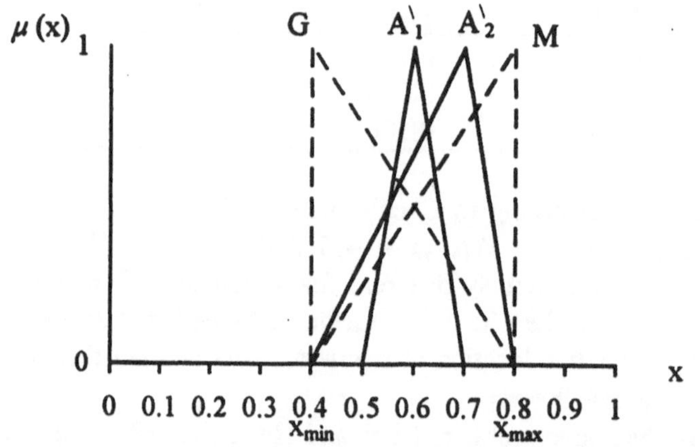

Figure 6.19b An example of the illogicality in Chen's method

Source: Chen et al., 1992, p. 246.

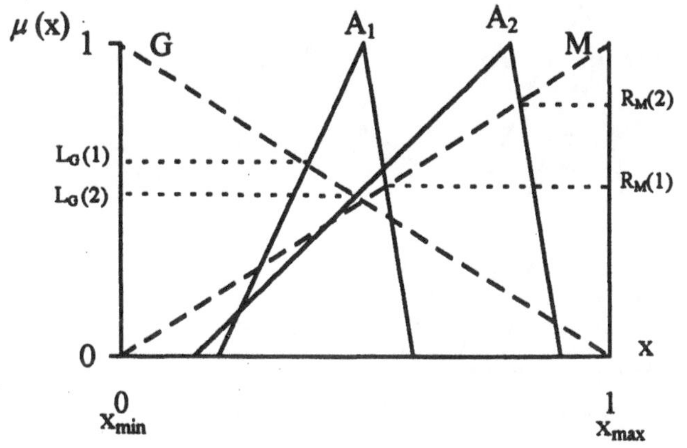

Figure 6.20 Chen and Hwang's method

Source: Chen et al., 1992, p. 247.

$$\mu_M(x) = \begin{cases} x & 0 \le x \le 1 \\ 0 & \textit{otherwise} \end{cases} \tag{6.95}$$

$$\mu_G(x) = \begin{cases} 1-x & 0 \le x \le 1 \\ 0 & \textit{otherwise} \end{cases} \tag{6.96}$$

where x, μ_M(x) and μ_G(x) take values in the range [0, 1].

The maximising and minimising sets in Chen and Hwang's method are defined so that the absolute locations of fuzzy numbers can be automatically incorporated in the comparison process. Computational procedures are the same as for Chen's method except that the definitions of μ_M(x) and μ_G(x) are different. An example is presented in figure 6.21 where the total scores for the three fuzzy numbers, A_1, A_2, and A_3 are .35, .37, and .44 respectively. Therefore, $A_3 > A_2 > A_1$ (Chen et al., 1992, pp. 248-249).

The advantage of Chen and Hwang's method over Chen's method is that it gives a crisp number (the total score) for each fuzzy number, and the

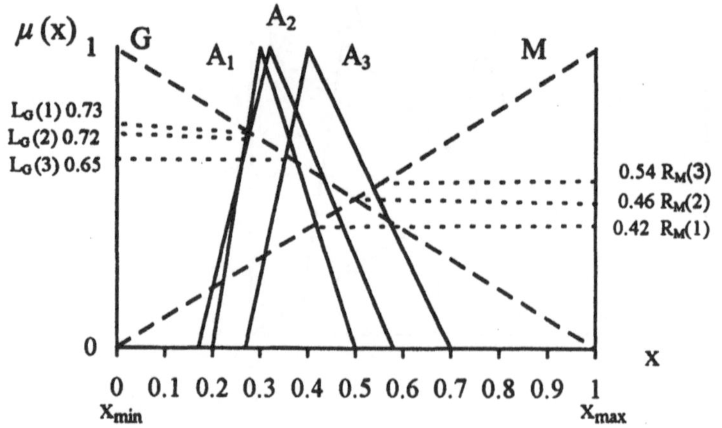

Figure 6.21 An example of Chen and Hwang's method

Source: Chen et al., 1992, p. 248.

crisp number takes account of the absolute location of the fuzzy number. However, the method restricts the fuzzy numbers in question to the closed interval [0,1]. Thus, if the fuzzy numbers considered are on a real line, $\Re$ $[-\infty, +\infty]$, it is not possible to define the maximising and minimising sets as Chen and Hwang suggest.

A Proposed Ranking Method

As already noted, Chen's method reflects both the right- and left-hand sides of the membership function of a fuzzy number in the ranking process. This means that the critical, 'scoring' members of the fuzzy number are those with a membership function less than 1. Thus, an important part of a fuzzy number is neglected in the ranking process, that is, the part of the fuzzy number with membership function equal to 1. In order to obtain a ranking method for fuzzy numbers which reflects *all* the important information contained in a fuzzy number, a new method is proposed.

Let A_i , i =1,2 ... n be fuzzy numbers on a real line, $\Re$, with membership functions $\mu_{Ai}(x)$. To accommodate both sides of the membership

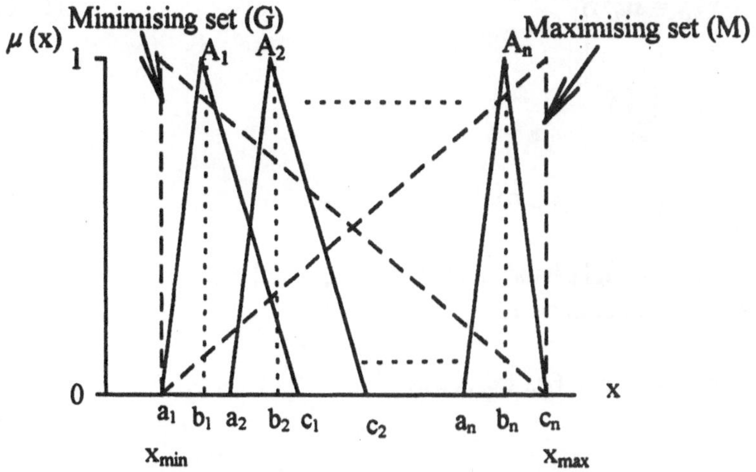

Figure 6.22 A proposed ranking method

function of the fuzzy numbers under consideration, the maximising set (M) and minimising set (G) are defined as follows (see figure 6.22):

$$M = \{x, \mu_M(x) \mid x \in \Re\} \tag{6.97}$$

$$G = \{x, \mu_G(x) \mid x \in \Re\} \tag{6.98}$$

where:

$$\mu_M(x) = \begin{cases} (x - x_{\min})/(x_{\max} - x_{\min}) & x_{\min} \le x \le x_{\max} \\ 0 & otherwise \end{cases} \tag{6.99}$$

$$\mu_G(x) = \begin{cases} (x_{\max} - x)/(x_{\max} - x_{\min}) & x_{\min} \le x \le x_{\max} \\ 0 & otherwise \end{cases} \tag{6.100}$$

and:

$$x_{min} = inf\, S \tag{6.101}$$

$$x_{max} = sup\ S \quad (6.102)$$

$$S = \bigcup_{i=1}^{n} S_i \quad (6.103)$$

$$S_i = \{x \mid \mu_{Ai}(x) > 0\} \qquad i = 1, \ldots n \quad (6.104)$$

The scores of right-hand side $R_M(A_i)$ and left-hand side $L_G(A_i)$ of a fuzzy number A_i are defined as:

$$R_M(A_i) = \sup(\mu_{Ai}(x) \wedge \mu_M(x)) \quad (6.105)$$

and:

$$L_G(A_i) = \sup(\mu_{Ai}(x) \wedge \mu_G(x)) \quad (6.106)$$

where $\wedge$ stands for minimum operation and i= 1, 2.....n.

Then, the total score of a fuzzy number A_i, $T(A_i)$, is defined as:

$$T(A_i) = (k)[R_M(A_i)] + (1\text{-}k)[1 - L_G(A_i)] \quad (6.107)$$

where i=1,...n, $R_M(A_i) \in [0,1]$, $L_G(A_i) \in [0,1]$, and $k \in [0,1]$.

The right and left-hand sides of the membership function of a fuzzy number represent the optimistic and pessimistic views respectively (Kim and Park, 1990). In equation 6.107, k represents an index of optimism, a real number in the closed interval [0,1] where '0' reflects the pessimistic view of the decision maker and '1' reflects the optimistic view of the decision maker[14].

Finally, the ranking value of a fuzzy number A_i, $V(A_i)$ is defined as:

$$V(A_i) = m_i \, . \, T(A_i) \quad (6.108)$$

where:

$$m_i = \max\{x \mid \mu_{Ai}(x) = 1\} \quad (6.109)$$

$V(A_i)$ represents a crisp number on the real line, $\Re$, associated with fuzzy number A_i. It might be interpreted as the member of the fuzzy number based on the highest membership function(s) and weighted by the average score of both sides of the membership function of the fuzzy number. Fuzzy numbers are ranked according to $V(A_i)$ and these values $[V(A_i)]$ reflect the relative locations of the fuzzy numbers A_i as well as their absolute locations on $\Re$.

In the case of two or more fuzzy numbers with the same $V(A_i)$, the divergence can be used to discriminate between them. The lowest divergence is the higher ranking number. If A_i is a triangular fuzzy number represented by three parameters (a_i, b_i, c_i), where $a_i \leq b_i \leq c_i$, then, the divergence δ_i of a fuzzy number A_i is defined as:

$$\delta_i = c_i - a_i \tag{6.110}$$

Under the proposed method, the ranking of fuzzy numbers A_i, A_j is based on the following rules:

- If $V(A_i) > V(A_j)$ then $A_i > A_j$
- If $V(A_i) = V(A_j)$ and $\delta_i < \delta_j$ then $A_i > A_j$
- If $V(A_i) = V(A_j)$ and $\delta_i = \delta_j$ then $A_i = A_j$

In the case of a triangular fuzzy number A_i defined as above with a membership function as in (6.111) and using the definition of $\mu_M(x)$ and $\mu_G(x)$ as in (6.99) and (6.100), the $V(A_i)$ can be obtained mathematically as follows:

$$\mu_{A_i}(x) = \begin{cases} (x - a_i)/(b_i - a_i) & a_i \leq x \leq b_i \\ (c_i - x)/(c_i - b_i) & b_i \leq x \leq c_i \\ 0 & otherwise \end{cases} \tag{6.111}$$

- $R_M(A_i) = \text{Sup}\,(\mu_{Ai}(x) \wedge \mu_M(x))$
  $= (c_i - x_{min}) / [(x_{max} - x_{min}) + (c_i - b_i)]$ (6.112)
- $L_G(A_i) = \text{Sup}\,(\mu_{Ai}(x) \wedge \mu_G(x))$
  $= (x_{max} - a_i) / [(x_{max} - x_{min}) + (b_i - a_i)]$ (6.113)

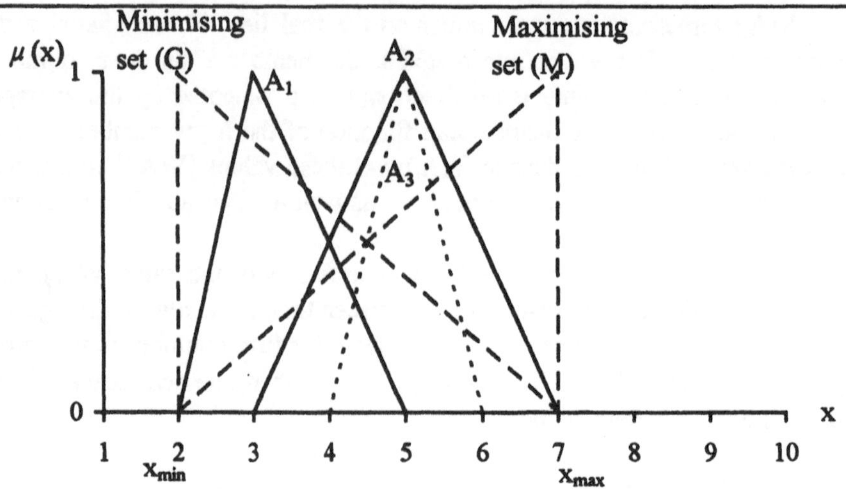

Figure 6.23 An example of the proposed ranking method

- $T(A_i) = (k)[R_M(A_i)] + (k-1)[(1 - L_G(A_i))]$
  $= (k)[(c_i-x_{min})/(x_{max}-x_{min}+c_i- b_i)]$
  $+(1-k)[1-(x_{max} - a_i)/(x_{max} - x_{min}+b_i-a_i)]$ (6.114)

- $V(A_i) = (m_i)((k)[(c_i-x_{min})/(x_{max} -x_{min}+c_i-b_i)]$
  $+ (1-k)[1-(x_{max} - a_i)/(x_{max}-x_{min}+b_i-a_i)])$ (6.115)

As an example to clarify the proposed ranking method, suppose there are three triangular fuzzy numbers (figure 6.23) defined as follows:

$A_1 = (2, 3, 5)$,
$A_2 = (3, 5, 7)$,
$A_3 = (4, 5, 6)$.

The ranking process for these numbers can be carried out as follows (assume k = 0.5):

$$S = \bigcup_{i=1}^{n} S_i$$
$$= \{2, 3, 4, 5, 6, 7\}$$

$x_{max} = sup\ S = 7$

$x_{min} = inf\ S = 2$

$V(A_i) = (m_i)((k)[(c_i-x_{min})/(x_{max}-x_{min}+c_i-b_i)]+(1-k)[1-(x_{max}-a_i)/(x_{max}-x_{min}+b_i-a_i)])$
$V(A_1) = (3)(1/2[(5-2)/(7-2+5-3)])+[1/2(1-((7-2)/(7-2+3-2))]) = 0.9$
$V(A_2) = (5)(1/2[(7-2)/(7-2+7-5)]+[1/2(1-((7-3)/(7-2+5-3))]) = 2.275$
$V(A_3) = (5)(1/2[(6-2)/(7-2+6-5)]+[1/2(1-((7-4)/(7-2+5-4))]) = 2.925$

∴ The ranking of the three numbers is $A_3 > A_2 > A_1$

As can be seen from the above example, the proposed ranking method is intuitively consistent. Although A_2 and A_3 share the same peak ($b_2 = b_3 = 5$), A_3 is preferred -because A_2's divergence [3, 7] is greater than A_3's [4, 6].

It might be noted that although the proposed ranking method is similar to Chen's method in defining the maximising and the minimising sets, there are three basic differences between the two ranking methods:

1. The proposed method gives a crisp number on the real line, $\Re$, whereas Chen's method gives a score for each fuzzy number in question. The crisp number in the proposed method reflects all of the important information in the fuzzy number: both sides of its membership function and the number(s) having full membership.
2. The proposed method uses a second criterion to distinguish between fuzzy numbers which have the same crisp number: the divergence of a fuzzy number. This is consistent with intuition that one prefers numbers with less divergence as they have greater precision.
3. The 'k' used in the proposed method is an index of optimism while the exponent 'k' in Chen's method seems meaningless in the context of decision analysis as there is no rationale for deriving any particular value.

Linguistic Variables

The linguistic variable approach was developed by Zadeh (1972; 1973; 1975a; 1975b; and 1975c) to provide a systematic means for characterising complex or ill-defined phenomena. He stated that:

'... by moving away from the use of quantified variables and toward the use of type of linguistic description employed by humans, we acquire a capability to deal with systems which are much too complex to be susceptible to analysis in conventional mathematical terms'. (Zadeh, 1973, p. 29)

A linguistic variable is defined by Zadeh (1973, p. 28) as 'a variable whose values are sentences in a natural or artificial language'. In other words, a linguistic variable takes values which are not numbers but are words or sentences. As an example, 'high' is a linguistic variable if its values are 'tall', 'very tall', 'not tall', etc. Each value is defined by a fuzzy subset in which the membership function represents the compatibility of each element in the universe of discourse (base values), with the label of the fuzzy subset.

More specifically, a linguistic variable is characterised by a quadruple (X, T(x), U, M(x)) (Liu and Wang, 1994, p. 467) where:

X is the name of the variable,
T(x) is the term-set of x, i.e., the collection of its linguistic values,
U is the universe of discourse, and
M(x) denotes the meaning of the linguistic value x, which is a fuzzy subset of U, and is characterised by a membership function $\mu_x\colon U \to [0,1]$.

As an example, Zadeh (1975a, p. 202) suggested that a linguistic variable called 'age', i.e. X = age, might have values 'young', 'very young', and 'old', i.e.,

T (age) = young + very young + old

In the universe of discourse, for example, U = [0, 100], the meaning of the linguistic variable 'old' might defined as:

M (old) = {(50, .6), (55, .8), (60, .9), (65, 1)}

To clarify the concept of a linguistic variable, figure (6.24) shows the hierarchical structure of the relation between the four components of the linguistic variable 'age' (Zadeh, 1975a)

It should be noted that the base values, the universe of discourse, can be numerical in nature, as above in the linguistic variable 'age'. Thus, it is

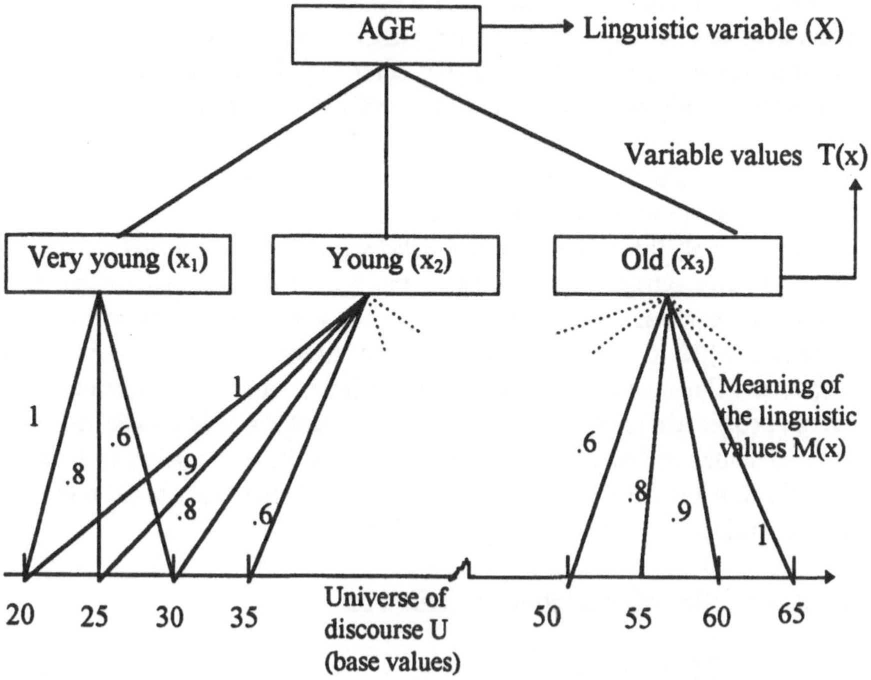

Figure 6.24 Hierarchical structure of a linguistic variable 'age'

Source: Zadeh, 1975a, p. 204.

possible to define the meaning of a linguistic value such as 'young', by a membership function which assigns each age in the interval [0,100] a number in the interval [0,1], and which represents the compatibility of that age with the label 'young'. On the other hand, the base values could be non-numerical such as the linguistic variable 'appearance' which does not have well-defined base values. In this case the membership function could be based on impression (Zadeh, 1975a).

In practice, if the base values are not numerical in nature, it is still possible to build up a base scale. The choice of a base scale should conform to four basic criteria (Cooley and Hicks, 1983): it should be able to accurately reflect the meaning of the linguistic values; it should not change the base values associated with a particular linguistic value because of low or moderate judgement uncertainties; it should recognise strong judgement changes by appropriate movements along the base scale; and it should not significantly change the results of a model using the linguistic values because of small changes in judgement.

In general, the value of a linguistic variable consists of two types of term: primary terms which represent the measurement values used in the evaluation, such as 'good', 'fair', and 'bad'; and hedge terms which act as modifiers when used in conjunction to primary terms to generate other linguistic values, such as 'very', 'more-or-less', 'slightly', 'extremely', etc. When hedges are joined with primary terms, additional linguistic values are generated such as 'very good', 'more-or-less good', 'extremely good', etc.

To deal mathematically with linguistic variables, two basic operations were introduced by Zadeh (1975a and 1975b):

1. The operation of concentration, defined as:

$$\text{CON (A)} = \text{A}^2 \tag{6.116}$$

Applying this operation to a fuzzy set 'A' results in a fuzzy set 'A\'' such that the reduction in the magnitude of the membership function of 'x' in 'A' is relatively small for those 'x' which have a high membership function in 'A' and relatively large for 'x' with a low membership function.

2. The operation of dilation, defined as:

$$\text{DIL (A)} = \text{A}^{.5} \tag{6.117}$$

The effect of this operation is opposite to that of concentration.

As an example, let U = {1,2, ... 10} and A = {(3,.4), (4,.6), (5,1), (6,.3)}, then CON (A) and DIL (A) are computed as follows:

$$\begin{aligned}\text{CON (A)} &= \text{A}^2\\ &= \{(3, .16), (4, .36), (5, 1), (6, .09)\}\end{aligned}$$

DIL (A) $= A^{.5}$
$= \{(3, .63), (4, .77), (5, 1), (6, .55)\}$

The operations of concentration and dilation are very important for deriving new linguistic values. As suggested by Zadeh (1972; 1975a; and 1975b), the concentration operation is used to derive the hedge 'very' and the dilation operation is used to derive the hedge 'more-or-less'[15]. To clarify this idea, let U = {1, 2, 3, 4, 5}, and the linguistic variable 'bad' defined, subjectively, as follows:

Bad = {(1, 1), (2, .8), (3, .6), (4, .4), (5, .2)}

Then:

Very bad $= (\text{bad})^2$
$= \{(1,1), (2, .64), (3, .36), (4, .16), (5, .04)\}$

and:

More-or-less bad $= (\text{bad})^{.5}$
$= \{(1, 1), (2, .9), (3, .77), (4, 63), (5, .45)\}$

In addition to the previous two operations, using the connectives, 'and' and 'or', and the negation, 'not', can generate additional linguistic values. Table 6.7 shows how additional linguistic values can be generated assuming that the meaning of the primary values 'good', 'fair', and 'bad' are already defined by their membership functions μ_G, μ_F, and μ_B respectively.

As an alternative to the operations of concentration and dilation in generating additional linguistic values, the decision analyst can define a specific base scale for all linguistic values used in the analysis of a particular situation. Chen et al. (1992, pp. 466-473) derived eight separate conversion scales using a base scale in the closed interval [0, 1] to convert linguistic values to fuzzy numbers. (see figures 6.25a - 6.25h).

It should be noted that, although a linguistic value such as 'high' may be used in many scales, the fuzzy numbers associated with it could be quite different from one scale to another. This reflects the fact that the same linguistic value may possess different meanings when used in different circumstances (Chen et al., 1992, p. 466).

Table 6.7 Generating additional linguistic values

Linguistic values	Membership function
Extremely Good	$\mu_{EG} = (\mu_G)^3$
Very Good	$\mu_{VG} = (\mu_G)^2$
More or Less Good	$\mu_{MG} = (\mu_G)^{.5}$
Not Good	$\mu_{NG} = 1 - \mu_G$
Not Good and Not Fair	$\mu_{NGF} = \min [(1 - \mu_G), (1 - \mu_F)]$
Good or Fair	$\mu_{GOF} = \max (\mu_G, \mu_F)$
Bad or Fair	$\mu_{BOF} = \max (\mu_B, \mu_F)$
Not Bad and Not Fair	$\mu_{NBF} = \min [(1 - \mu_B), (1 - \mu_F)]$
More or Less Bad	$\mu_{MB} = (\mu_B)^{.5}$
Very Bad	$\mu_{VB} = (\mu_B)^2$
Extremely Bad	$\mu_{EX} = (\mu_B)^3$

Source: Wang et al., 1989, p. 200.

A summary of the linguistic values used in the eight scales is shown in table 6.8. The principle involved in choosing a specific scale is to pick a figure that contains all of the linguistic values used by a decision maker. The fuzzy numbers in that figure can then be used to represent the meaning of the linguistic values. For instance, if the decision maker uses the terms: 'medium' and 'very high' then figure 6.25c, scale 3, should be used because this figure (scale) contains the terms 'medium' and 'very high'. Alternatively, if the terms 'medium' and 'high' are used by a decision maker then although all the scales contain these two terms, the simplest scale (figure 6.25a) should be chosen (Chen et al., 1992, p. 470).

It is worth noting that the linguistic values used in the eight scales are in the universe U = {excellent, very high, high to very high, high, fairly high, medium, fairly low, low, low to very low, very low, none}. This universe is appropriate when discussing certain phenomena (e.g. the quality of production), however, it is not suitable for discussing other phenomena, such as, for example, the distance between two places or two objects. The universe must therefore be modified to fit the nature of attributes of a specific decision problem. For example, if size was one of the attributes, then the primary terms 'high' and 'low' in the above universe should be replaced by 'large' and 'small' respectively.

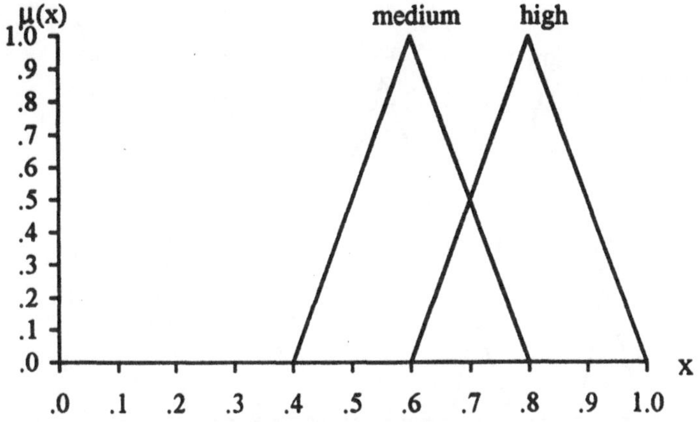

Figure 6.25a Scale 1

Source: Chen et al., 1992, p. 467.

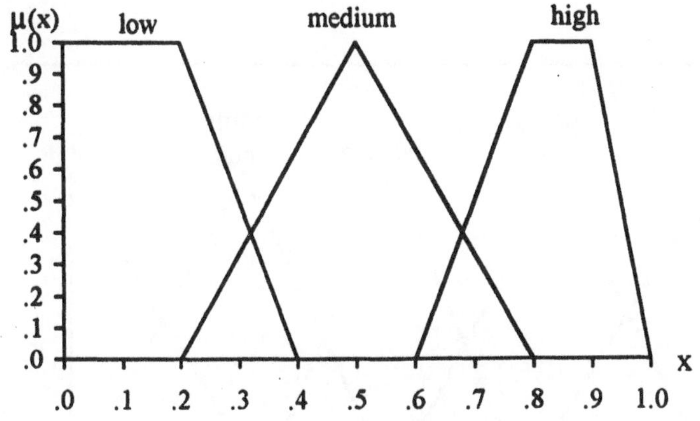

Figure 6.25b Scale 2

Source: Chen et al., 1992, p. 467.

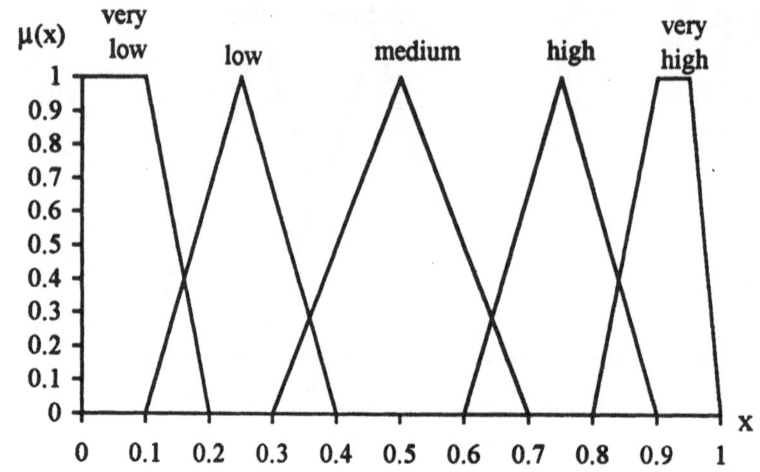

Figure 6.25c Scale 3

Source: Chen et al., 1993, p. 467.

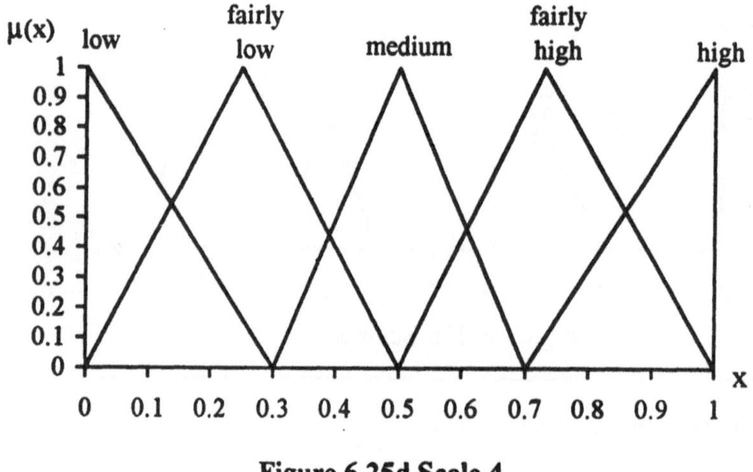

Figure 6.25d Scale 4

Source: Chen et al., 1992, p. 468.

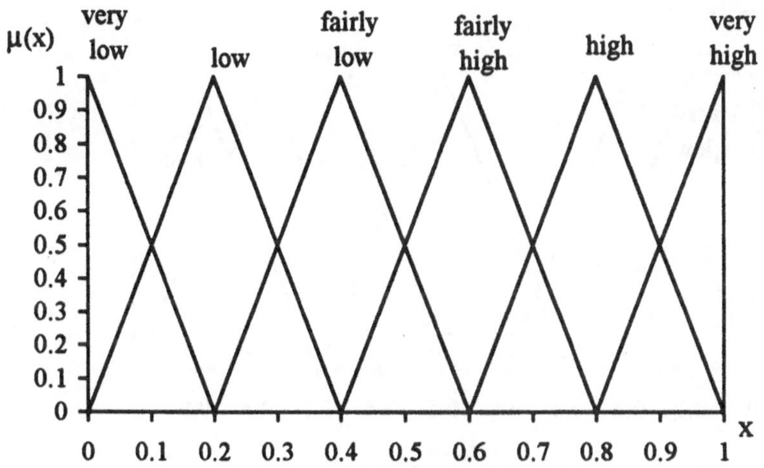

Figure 6.25e Scale 5

Source: Chen et al., 1992, p. 468.

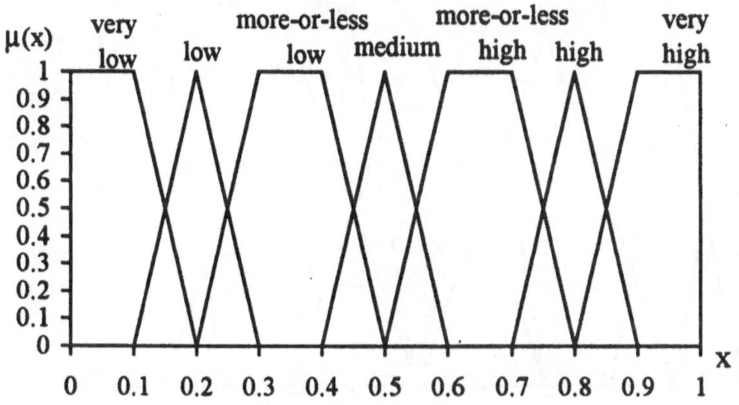

Figure 6.25f Scale 6

Source: Chen et al., 1992, p. 468.

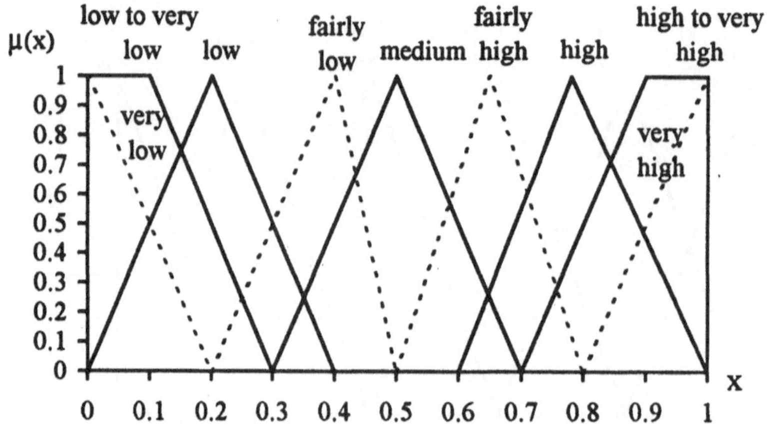

Figure 6.25g Scale 7

Source: Chen et al., 1992, p. 469.

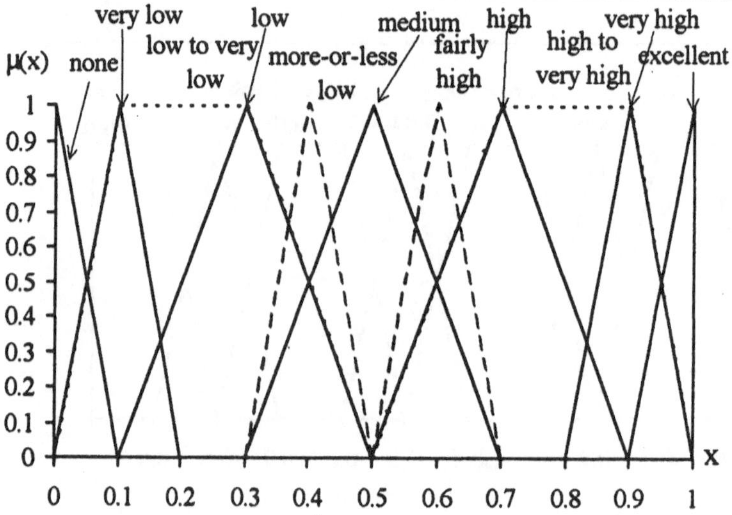

Figure 6.25h Scale 8

Source: Chen et al., 1992, p. 469.

Table 6.8 Summary of linguistic values used in the eight scales

Scale	1	2	3	4	5	6	7	8
No. of terms used	two	three	five	five	six	seven	nine	eleven
None								✓
Very low			✓		✓	✓	✓	✓
Low to very low							✓	✓
Low		✓	✓	✓	✓	✓	✓	✓
Fairly low				✓	✓		✓	
More or less low						✓		✓
Medium	✓	✓	✓	✓		✓	✓	✓
More or less high						✓		
Fairly high				✓	✓		✓	✓
High	✓	✓	✓	✓	✓	✓	✓	✓
High to very high							✓	✓
Very high			✓		✓	✓	✓	✓
Excellent								✓

Source: Chen et al, 1992, p. 470.

Conclusions

Fuzzy set theory was developed as a device to deal mathematically with imprecise and ambiguous expressions, and can be applied to both numerical and linguistic data. Before discussing these two elements, basic fuzzy set theory was introduced, fuzzy sets were distinguished from ordinary (crisp) sets and the main operations (complement, intersection, and union) were extended from ordinary sets to fuzzy sets.

To deal mathematically with fuzzy numbers two conditions should be satisfied: normality and convexity. Normality means that a fuzzy number should have at least one element which has full membership, convexity means that there is only one peak for a fuzzy number.

As fuzzy algebraic operations may involve extensive computations, special forms of fuzzy number have been proposed in the fuzzy set literature. These include L-R fuzzy numbers, L-R flat fuzzy numbers triangular fuzzy numbers, and trapezoidal fuzzy numbers. Basic fuzzy algebraic operations have been reviewed for each of these types of fuzzy number. By applying such fuzzy operations, many of the techniques of management science (such as linear programming, goal programming and game theory) can be extended to accommodate fuzzy data. This would overcome one of the criticisms of these techniques: that they require precise data, and are therefore difficult to apply to practical decision problems.

Using fuzzy numbers in decision making contexts leads to the 'ranking problem' if there are several alternatives and each alternative is presented as a fuzzy number. This occurs because each fuzzy number is represented by imprecise quantities, and so there may be an overlap between these quantities across the fuzzy numbers. This overlapping can thus lead to difficulty in ranking fuzzy numbers consistently. A review of existing ranking methods shows that different methods can lead to different orders for the same fuzzy numbers. Different methods have been criticised on various grounds, they: may lead to counter-intuitive results; consider only the relative locations of fuzzy numbers, ignoring their absolute locations; and/or ignore an important part of the fuzzy numbers in the ranking process (that is, those elements with membership functions equal 1). A ranking method which overcomes these criticisms and reflects all the important information in the fuzzy numbers was proposed in this chapter.

The proposed ranking method is based on two criteria for ranking fuzzy numbers. First, it provides a ranking value for each fuzzy number. This ranking value is derived by weighting those members of the fuzzy number with the highest membership functions by the average score of both sides of the membership function of the fuzzy number. However, if two fuzzy numbers have the same ranking value, a second criterion, the divergence, can be applied in order to differentiate them. If two or more fuzzy numbers cannot be distinguished by these two criteria, then they are equal.

The fuzzy linguistic approach is based on the concept of a linguistic variable. The essential feature of a linguistic variable is that its values take a linguistic form (word or sentence) and not a number. Each linguistic value is represented by a fuzzy subset in which the membership function represents the compatibility of each element, in the universe of discourse, with the label of the fuzzy subset. Thus, linguistic values can be dealt with mathematically using fuzzy subsets as surrogates.

Recently, fuzzy set theory has been used in solving accounting problems such as cost/benefit analysis (Wang and Liang, 1995), cost-volume-profit analysis (Chan and Yuan, 1990), evaluation of internal control (Cooley and Hicks, 1983), cost variance investigations (Zebda, 1984; and Lin and O'Leary, 1989), audit sampling (Lin, 1984), depreciation (Rhys, no date), financial ratio analysis (Gutierrez and Carmona, 1988) and stock selection (Hammerbacher and Yager, 1981). The suitability of fuzzy set theory in accounting decision problems has been examined by, amongst others, Zebda (1989 and 1991) and de Korvin (1995), and four reasons have been suggested for the application of the approach to accounting problems. First, fuzzy set theory provides a mathematical framework wherein fuzzy accounting terms and concepts (e.g., 'high profit', 'efficient sampling procedure', and 'material error') can be treated systematically. Thus, by using fuzzy set theory, accountants no longer need to ignore the ambiguity in accounting or treat it as randomness to be analysed by using the probability theory. An examination of statements, terms, and rules frequently used by accountants reveals that ambiguity exists in many accounting problems. For example, 'the financial statements present *fairly* the financial position of the firm', 'the internal control system is *strong*', 'if there are *material* weakness in the internal control, expand sample size', 'investigate *large* or *significant* variances', and 'joint products are classified as by-products, if they have *low* selling prices'. Such statements, terms, and rules involve ambiguity because of the imprecise meaning of the italicised words (Zebda, 1991).

Second, fuzzy set theory reduces the need for precise inputs in decision analysis by allowing the use of imprecise estimates of the values of variables, the relationships between variables, and the values of probabilities. For example, the estimates of a variable may be expressed in terms such as 'high', 'low', 'medium', 'between £2000 and £3000', 'about 30 percent', or 'quite likely'. Reducing the need for precise measures is likely to increase the relevance of accounting analysis because relevant items do not need to be ignored just because they do not have precise measurements (de Korvin, 1995).

Third, as fuzzy set theory relaxes the 'law of the excluded middle' and 'two-value logic', it reduces the need for binary classifications of accounting objects that are often artificial and unrealistic. For example, Zebda (1984) noted that in cost variance investigation, classifying performance as either in-control or out-of-control is unrealistic.

Finally, unlike traditional quantitative methods, fuzzy set theory allows for varying the degree of precision. Fuzzy set theory allows for the use of imprecise as well as precise estimates of accounting variables and relationships. For example, estimates of cost in a decision model can be expressed vaguely, as when one uses values such as 'high profits'. Precise estimates such as '£1000' can be described by a fuzzy set $C = \{1000, 1\}$. Thus, by using fuzzy set theory, decision analysts can express their perceptions of the imprecision of the required data and consequently it is no longer necessary to (mistakenly) assume a constant (high) level of precision (Zebda, 1989).

The development of successful applications of fuzzy set theory to many accounting problems (see, Siegel et al., 1995) provides encouragement that it will be similarly appropriate in AMT investment decisions. These decisions can be affected by a wide range of factors; some of these factors are qualitative (such as process flexibility, product quality, and customer requirements) and others are quantitative (such as NPV and payback period). Fuzzy linguistic variables provide a means of handling qualitative factors while any imprecision in the quantitative data can be appropriately modelled using fuzzy (rather than crisp) numbers.

In many practical situations, analysts deliberately utilise only a fraction of the information available in the initial data, ignoring all information which relates to imprecision (Pohjola and Turunen, 1990). The source of imprecision can be related either to the uncertainty of future events or to ambiguity/vagueness (Franz et al., 1995). When imprecision is related to uncertainty, it is usually handled by probability theory (Bellman and Zadeh, 1970). However, probability theory also requires precise estimates for the probabilities of the data (Zebda, 1989). Moreover, the empirical research reported in the previous two chapters indicates that probability theory is rarely used in practice. If imprecision is related to ambiguity or vagueness, then fuzzy set theory can be used to handle such imprecision. As will be seen in the next chapter, when modelling AMT investment decisions, a distinction is made between the uncertainty (riskiness) of a project and imprecision (ambiguity and/or vagueness) in estimating the required data.

Notes

1 Zebda (1989) and Kandel (1986), among others, distinguished between ambiguity and vagueness. Ambiguity exists when a word or concept has multiple meanings and is used to describe distinguishable subconcepts, while vagueness exists when the word or concept lacks precise, sharp boundaries. However, both terms imply imprecision and inexactness which are the thrust for using fuzzy set theory.

2 This chapter is not intended to provide full details for all aspects of fuzzy set theory. Rather, only concepts relating to the formulation of the model for evaluating AMT are discussed here. For further details, see for example: Zadeh (1965,1986, 1972, 1973, 1975a, b, and c), Gluss (1973), Baldwin and

Pilsworth (1982), Okuda et al. (1987), Dubois and Prade (1980 and 1989), Zimmermann (1987, 1991), Zadeh et al. (1975), and Kruse et al. (1994).

3 Existing ranking methods described in the literature utilise only partial information from each fuzzy number, and leave the decision maker to choose between 'optimistic' views (ranking based on right-hand side of the fuzzy numbers) or 'pessimistic' views (ranking based on the left-hand side of the fuzzy numbers). The proposed method utilises all the information in the fuzzy numbers to be ranked and allows the decision maker to weight the ranking value according to his or her degree of optimism/pessimism. See pp. 214-219.

4 For more details about classical or ordinary (non-fuzzy) set theory, see, for example: Growar and Flegg (1973), Rueff and Jeger (1970), and Hrabacek and Jech (1984).

5 $\forall$ means 'for all' and $\in$ means 'belonging to'.

6 The intersection and union could be modelled using operators other than min ($\wedge$) and max ($\vee$). These operators are chosen here because they are widely used in many applications (see, for example, Zebda, 1989). For more details about other operators see, for example: Chen et al. (1992), Zimmermann (1991), and Smithson (1987).

7 See pp. 183-184.

8 The Cartesian product of two ordinary sets A $\times$ B is an ordinary set of elements which are ordered pairs of each element in A and each element in B. For example, if A = {4, 5, 6} and B = {2, 3}, then A $\times$ B = {(4, 2), (4, 3), (5, 2), (5, 3), (6, 2), (6, 3)}. In fuzzy set theory the Cartesian product of A_1, ... A_n is defined as: $C = \int (x_1, \ldots . x_n) / \min (\mu_{A1}(x_1) \ldots .. \mu_{An}(x_n))$. For example, if A_1 = {(2, .3), (4, .5)} and A_2 = {(1, 1), (2, .4)}, then, $A_1 \times A_2$ = {[(2; 1), .3], [(2; 2), .3], [(4; 1), .5], [(4; 2), .4]}.

9 See for example: Zimmermann, 1991, p. 53; Dubois and Prade, 1980, p. 14; Pal and Dutta Majumeder, 1987, p. 46; Kandel, 1986, p. 15; and Chen et al., 1992, pp. 64-65.

10 To distinguish algebraic fuzzy operations from normal algebraic operations, the following signs are used (+), (−), (•), and (+) for fuzzy addition, fuzzy subtraction, fuzzy multiplication, and fuzzy division respectively.

11 The mean value is that value which has membership function equal to 1.

12 Chen et al. (1992, p. 239) referred to k as an indicator of the decisions maker's attitude toward risk; when some $k < 1$ is chosen, the decision maker is said to have a risk-averse attitude, whereas $k > 1$ is chosen when the decision maker has a risk-taking attitude.

13 It was not possible to obtain the original unpublished Chen and Hwang (1989) paper. Thus all the discussion is based on Chen et al. (1992, pp. 251-264).

14 In equation 6.107, if k = 1, this gives a value equal to Jain's ranking value while if k = 0.5, Chen's ranking value is derived. The proposed method is therefore more general and allows degrees of optimism or pessimism to be made explicit.

15 Using the operations of concentration and dilation to refer to the modifiers 'very' and 'more-or-less' has been debated. A summary can be found in Smithson (1987, pp. 67- 72).

7 A Fuzzy Approach to the Evaluation of AMT Investment Projects

The review of the literature relating to the evaluation of AMT investment projects indicated that the main difficulty was quantification of the qualitative factors so that they could be incorporated into the decision model. This difficulty was confirmed in the empirical study. Eighty nine percent of respondents who had experience in evaluating AMT rated 'quantifying intangible factors' as a difficult/extremely difficult task in the evaluation process. It was also found that numerous theoretical approaches have been developed in order to overcome this problem. These approaches range from denying that the problem exists and claiming that all factors could be quantified in cash flow (for example, Primrose, 1991), to recommending that the investment in AMT be regarded as a strategy that should be implemented regardless of the results of cash flow analysis. Between these two extremes, a number of researchers have suggested models based on analytical techniques which aim to quantify all of the factors of AMT investments. These models were reviewed in chapter 2. However, in general, these models have a common deficiency, emphasising analytical elegance rather than empirical validity. In each model, there is insufficient justification for the inclusion of some factors and not others. Overcoming this deficiency was a key motivation for the empirical study reported in chapters 4 and 5 as the authors believe that decision makers will be more convinced by a model based on practical experience than on prescription.

The examination of practice in a sample of large UK manufacturing companies confirmed the problem. All the companies sampled based their evaluation not only on financial analysis but also with considerable attention to non-financial factors. A criticism that can be levelled against the empirical practices described in chapter 5 is their lack of a consistent method for accounting for non-financial factors. In some cases these factors are merely discussed as supportive of a decision based mainly on financial analysis. It is therefore proposed that such 'qualitative factors' be taken into account by extending the traditional two-dimensional 'return-risk' financial analysis to include a third dimension for the so-called 'intangible benefits' when evaluating AMT investment projects.

In order to make AMT decisions in an objective and consistent manner, there is a need for a device that can consistently treat each group of factors, whether quantitative or qualitative. The approach suggested here is the use of fuzzy set theory, which permits the development of a model that can handle linguistic (verbal) variables as well as variables expressed as approximate, rather than precise, numerical data. The model developed in this chapter is based on fuzzy set theory and thus provides a means of combining qualitative and quantitative variables in the investment decision.

The chapter is organised as follows:

1. A review of previous studies using fuzzy set theory for the evaluation of capital investment projects.
2. A proposed model for AMT investment decisions.
3. An illustrative example.
4. Conclusions.

A Review of Previous Studies Using Fuzzy Set Theory for the Evaluation of Capital Investment Projects

Two approaches have been proposed in the literature that use the concepts of fuzzy set theory for the evaluation of capital investment projects. The first approach is based on the concept of fuzzy numbers and extends traditional discounted cash flow analysis into fuzzy cash flow analysis (Ward, 1985 and 1989; and Chiu and Park, 1994) while, the second approach is based on the concept of linguistic variables (Wilhelm and Parsaei, 1988 and 1991). Both approaches are reviewed below.

Ward (1985 and 1989) developed fuzzy present value analysis in which the cash flows are modelled as fuzzy numbers rather than crisp ones. He suggested that trapezoidal fuzzy numbers (flat fuzzy filter function numbers) be used in estimating the cash flows. The fuzzy present value could be defined as follows:

$$P(i) = A(0)\ (+)\ A(1)\ (\bullet)\ (1 + i)^{-1}\ (+) \ldots (+)\ A(N)\ (\bullet)\ (1 + i)^{-N} \qquad (7.1)$$

where:

P(i) is the fuzzy present value,
A(n) is a cash flow having fuzzy amount and flows at the end of period n (n = 0, 1, ..., N), and
i is the effective interest rate for end-of- period compounding.

This formula is a straightforward extension of the non-fuzzy (crisp) present value formula based on cash flows specified as fuzzy numbers. However, derivation of the membership function of fuzzy present value (P) requires tedious computational effort because, even though every cash flow amount has a linear membership function, it results in complex non-linear representations[1].

To overcome this problem and so make fuzzy cash flow analysis more manageable, Chiu and Park (1994) developed an approximate form of the 'complex' fuzzy present value formula. They based their suggestion on triangular fuzzy numbers instead of the trapezoidal numbers suggested by Ward (1985, 1989), defining a triangular fuzzy number 'P = (a, b, c)' with a linear representation on its left and right such that its membership function can be defined by using α notation (see chapter 6) as follows (see figure 7.1):

$$P = [P^{l(\alpha)}, P^{r(\alpha)}] = [a + (b - a)\alpha, c + (b - c)\alpha] \qquad \forall \alpha \in [0, 1] \qquad (7.2)$$

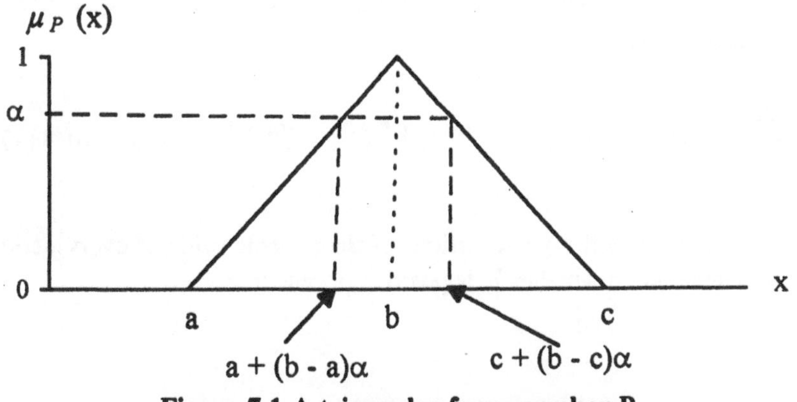

Figure 7.1 A triangular fuzzy number P

where α is the degree of membership function which maps to its corresponding x values on each side of P. The two values $P^{l(\alpha)}$ and $P^{r(\alpha)}$ are the x values in a linear representation of the left and right sides of P, to which each α maps, respectively.

Then, the periodic cash flow at time t can be represented by a triangular fuzzy number $P_t = (p_{t0}, p_{t1}, p_{t2})$ and characterised by its left and right representations at each degree of membership function α, as:

$$P_t = [p_t^{l(\alpha)}, p_t^{r(\alpha)}] = [p_{t0} + (p_{t1} - p_{t0})\alpha\, ,\, p_{t2} + (p_{t1} - p_{t2})\alpha\,] \quad \forall\, \alpha \in [0, 1] \quad (7.3)$$

P_t can be either disbursement ($p_t < 0$) or receipt ($p_t > 0$).

In the same way, the discount rate during time t can be defined by a triangular fuzzy number $R_t = (r_{t0}, r_{t1}, r_{t2})$, $\forall\, r > 0$. This fuzzy number can also be denoted by its left and right representations as:

$$R_t = [R_t^{l(\alpha)}, R_t^{r(\alpha)}] = [r_{t0} + (r_{t1} - r_{t0})\alpha, r_{t2} + (r_{t1} - r_{t2})\alpha\,] \ \forall\, \alpha \in [0, 1] \quad (7.4)$$

After defining the periodic cash flow and the discount rate, the present value of periodic cash flows is:

$$PV = \sum_{t=0}^{n} \frac{P_t}{\prod_{t'=0}^{t}(1+R_{t'})} \quad (7.5)$$

As R ≥ 0, which implies $(1+R_{t'})$ is positive, the right and left representations are:

$$\prod_{t'=0}^{t}(1+R_{t'}) = \left[\prod_{t'=0}^{t}(1+R_{t'}^{l(\alpha)}), \prod_{t'=0}^{t}(1+R_{t'}^{r(\alpha)})\right] \quad \forall\, \alpha \in [0, 1] \quad (7.6)$$

The right and left representations of the present value of every periodic cash flow depend on whether P_t is positive or negative.

If P_t is positive:

$$PV = \left[\frac{P_t^{l(\varepsilon)}}{\prod_{t'=0}^{t}(1+R_{t'}^{r(\alpha)})}, \frac{P_t^{r(\varepsilon)}}{\prod_{t'=0}^{t}(1+R_{t'}^{l(\alpha)})} \right] \quad \forall\, \alpha \in [0, 1] \quad (7.7)$$

If P_t is negative:

$$PV = \left[\frac{P_t^{l(\varepsilon)}}{\prod_{t'=0}^{t}(1+R_{t'}^{l(\alpha)})}, \frac{P_t^{r(\varepsilon)}}{\prod_{t'=0}^{t}(1+R_{t'}^{r(\alpha)})} \right] \quad \forall\, \alpha \in [0, 1] \quad (7.8)$$

The difference between the present value formula (7.5) and that of Ward (7.1) is that the discount rate (R) in (7.5) can be a fuzzy number, and that all the fuzzy numbers in (7.5) take triangular shape with linear representation.

Chiu and Park (1994) provide an illustrative example for the present value calculation of fuzzy project. In their example they derive the exact fuzzy present value and an approximate fuzzy present value, and then compare them in order to assess the reliability of the approximate fuzzy present value as a surrogate for the exact fuzzy present value.

They considered a three-year project with the estimated cash flow and interest rates specified by triangular fuzzy numbers as:

$P_0 = (-110, -100, -90)$
$P_1 = (-80, -60, -40)$ and $R_1 = (6\%, 7\%, 8\%)$
$P_2 = (110, 130, 140)$ $R_2 = (6\%, 7\%, 9\%)$
$P_3 = (100, 110, 130)$ $R_3 = (6\%, 8\%, 10\%)$

This information can be denoted by the left and right representations as:

$P_0 = [-110 + 10\alpha, -90 - 10\alpha]$
$P_1 = [-80 + 20\alpha, -40 - 20\alpha]$ and $1 + R_1 = [1.06 + .01\alpha, 1.08 - .01\alpha]$
$P_2 = [110 + 20\alpha, 140 - 10\alpha]$ $1 + R_2 = [1.06 + .01\alpha, 1.09 - .02\alpha]$

$P_3 = [100 + 10\alpha, 130 - 20\alpha]$ $\quad\quad$ $1 + R_3 = [1.06 + .02\alpha, 1.10 - .02\alpha]$

The exact fuzzy present value can be calculated by using the left and right representations as follows:

$$PV = [PV^{l(\alpha)}, PV^{r(\alpha)}]$$

where:

$$\begin{aligned} PV^{l(\alpha)} &= -110 + 10\alpha \\ &\quad + (-80 + 20\alpha) / (1.06 + .01\alpha) \\ &\quad + (110 + 20\alpha) / (1.08 - .01\alpha)(1.09 - .02\alpha) \\ &\quad + (100 + 10\alpha) / (1.08 - .01\alpha)(1.09 - .02\alpha)(1.1 - .02\alpha) \end{aligned}$$

$$\begin{aligned} PV^{r(\alpha)} &= -90 - 10\alpha \\ &\quad + (-40 - 20\alpha) / (1.06 + .01\alpha) \\ &\quad + (140 - 10\alpha) / (1.06 + .01\alpha)(1.06 + .01\alpha) \\ &\quad + (130 - 20\alpha) / (1.06 + .01\alpha)(1.06 + .01\alpha)(1.06 + .02\alpha) \end{aligned}$$

The approximate fuzzy present value (PVA) can be calculated as follows:

$$PVA = (PVA_0, PVA_1, PVA_3)$$

where:

$$\begin{aligned} PVA_0 &= -110 + [-80 /(1.08)] + [110 / (1.08)(1.09)] \\ &\quad + [100 / (1.08)(1.09)(1.1)] = -14.8048 \\ PVA_1 &= -100 + [-60 /(1.07)] + [130 / (1.07)(1.07)] \\ &\quad + [110 / (1.07)(1.07)(1.08)] = 46.4336 \\ PVA_2 &= -90 + [-40 /(1.08)] + [140 / (1.06)(1.06)] \\ &\quad + [130 / (1.06)(1.06)(1.06)] = 106.713 \end{aligned}$$

This result is a triangular fuzzy number which can be represented by the left and right representations as follows:

$$\begin{aligned} PVA &= [PVA^{l(\alpha)}, PVA^{r(\alpha)}] \\ &= [PVA_0 + (PVA_1 - PVA_0)\alpha, PVA_2 + (PVA_1 - PVA_2)\alpha] \\ &= [-14.8048 + 61.2384\alpha, 106.713 - 60.2794\alpha] \end{aligned}$$

The results of both the exact fuzzy present value (PV) and the approximate fuzzy present value (PVA) are depicted in figure 7.2. The most promising present value is 46.4336. The largest possible present value is 106.7 while the smallest possible present value is -14.8 (Chiu and Park, 1994).

Clearly, the three parameters of the fuzzy present value in both forms (exact and approximate) are (and should be) exactly the same, but they are different in their membership functions. The membership function of the exact fuzzy present value takes non-linear representations while the membership function of the approximate fuzzy value takes linear representations. To assess the reliability of the approximate fuzzy present value as a surrogate for the exact one, Chiu and Park (1994) followed Kaufmann and Gupta (1991, ch. 6) by measuring the maximum deviation between both present values.

The deviation between the exact present value (PV) and its approximate form (PVA) can be calculated on their left and right representations as d_l and d_r respectively:

$$d_l = PVA_0 + (PVA_1 - PVA_0)\alpha - PV^{l(\alpha)} \tag{7.9a}$$
$$d_r = PVA_2 + (PVA_1 - PVA_2)\alpha - PV^{r(\alpha)} \tag{7.9b}$$

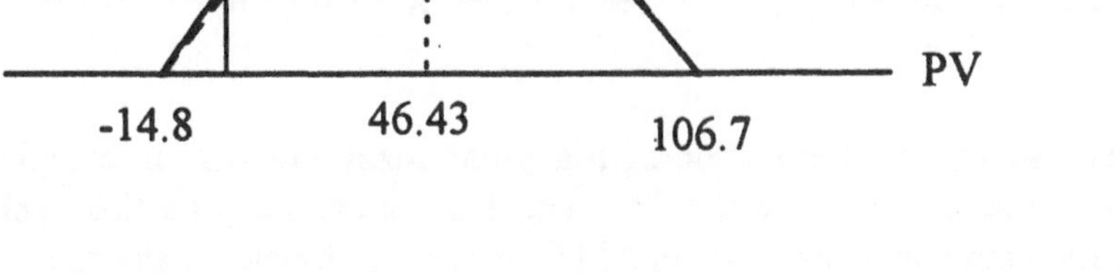

Figure 7.2 Exact fuzzy present value and its approximate form

Source: Chiu and Park (1994), p. 128.

The maximum deviation of both left and right representations can be found by taking very small increments in α, e.g., 0.001. Chiu and Park (1994) found the maximum deviation on the left PV representation, d_l^*, is at $\alpha = 0.508$ with a value of 0.211, and the maximum deviation on the right PV representation, d_r^*, is located at $\alpha = 0.499$ with a value of 0.1833. Then, these maximum deviations can be calculated as percentages of deviations, $d_{l(\%)}^*$ and $d_{r(\%)}^*$ for the left and right maximum percentage of deviations respectively, as follows:

$$d_{l(\%)}^* = d_l^* \;/\; (PVA_1 - PVA_0)\ \% \tag{7.10a}$$
$$d_{r(\%)}^* = d_r^* \;/\; (PVA_2 - PVA_1)\ \% \tag{7.10b}$$

Accordingly, $d_{l(\%)}^* = 0.345\%$ and $d_{r(\%)}^* = 0.304\%$. Because both percentages are 'negligibly small amounts' (less than 1%), Chiu and Park concluded that there is no significant deviation between PV and PVA, and this implies that PVA is a reliable approximate form of PV. Further, they examined which factors in the present value formula (periodic cash flows and discount rates) most affect the deviation. A computer-based-simulation was performed under different ranges of periodic cash flows and discount rates. Based on the results of the simulation, they concluded that:

1. The range of periodic cash flows does not significantly affect the deviation of PV and PVA.
2. The range of discount rates significantly affects the deviation of PV and PVA; the deviations increase significantly as the range of discount rates becomes larger.
3. However, when the confident width of discount rate is within a range of ±4%, the maximum deviation percentage is relatively small, which indicates that PVA can be regarded as a surrogate for PV in such circumstances.

The successful application of triangular fuzzy numbers in calculating present values as shown in the Chiu and Park study suggests that such an application may be appropriate in AMT investment decision making.

In general, it is important to note that the approximation does not affect the three parameters of the calculated triangular fuzzy number. The approximation simply permits the use of the linear representation of the triangular fuzzy number instead of using the non-linear representation which usually results from the operations of multiplication and division. As the simplification does not materially affect the final result of fuzzy analysis, it may be of more practical use.

Wilhelm and Parsaei (1988 and 1991) were the first to introduce the fuzzy linguistic approach to AMT investment decision making. They defined two fuzzy linguistic variables: X = '*IMPORTANCE*' and Y = '*CAPABILITY*'. The first variable is used to specify the *importance* associated with each strategic goal within a set of goals common to all alternative technologies, while the second variable is used to specify the *capability* of each technology to meet the strategic AMT goals. Each variable takes three primary values defined within the universe of discourse [0,1]. The three primary values for the variable '*IMPORTANCE*' are *critical*, *important*, and *unimportant*, while the values for the variable '*CAPABILITY*' are *superior*, *average*, and *poor*. Then, the fuzzy operators of negation or 'not'; the connectives 'and' and 'or'; and the linguistic hedges 'indeed', 'more-or-less', 'above', 'below', etc., together with the primary values, can generate new variable values. Table (7.1) summarises the membership functions for the values of the primary variables and their hedges[2].

Using these two fuzzy linguistic variables, Wilhelm and Parsaei (1988 and 1991) developed a heuristic algorithm to characterise the capability of alternative advanced technologies, and to select the best technology at each phase in the implementation of AMT strategy. They assumed that the company has a set of m common goals for AMT strategy, and that the importance of each goal should be assessed using the linguistic variable '*IMPORTANCE*'. The capability of each of the n available AMT projects should be assessed using the linguistic variable '*CAPABILITY*'. Then, for each combination of m goals and n technologies, the relations r_{ij} can be determined using the equation (7.11).

$$r_{ij} = x_i^{-1} (\alpha) y_j \tag{7.11}$$

where:

x_i^{-1} = the inverse of x,

(α) = a compositional operator defined such that the membership function of $x_i^{-1} (\alpha) y_j$ is determined from

$$f_{M(x-1(\alpha)y)} = \wedge[f^{-1}_{M(x)}(\mu_x) \ \alpha \ f_{M(y)}(\mu_y)],$$

Table 7.1 Membership functions for linguistic variables X (*IMPORTANCE*) and Y (*CAPABILITY*)

	Universe of discourse										
	0.0	**0.10**	**0.20**	**0.30**	**0.40**	**0.50**	**0.60**	**0.70**	**0.80**	**0.90**	**1.0**
Panel A: X variable values											
Indeed critical	0	0	0	0	0	0	0	0.01	0.05	0.92	1.0
Critical	0	0	0	0	0	0	0	0.05	0.15	0.80	1.0
More-or-less critical	0	0	0	0	0	0	0	0.22	0.39	0.89	1.0
Very important	0	0.01	0.06	0.56	0.81	1.0	0.81	0.56	0.06	0.01	0
Important	0	0.10	0.25	0.75	0.90	1.0	0.90	0.75	0.25	0.10	0
More-or-less important	0	0.32	0.50	0.87	0.95	1.0	0.95	0.87	0.50	0.32	0
Unimportant	1.0	0.80	0.40	0.20	0.05	0	0	0	0	0	0
Not important	1.0	0.90	0.75	0.25	0.10	0	0.10	0.25	0.75	0.90	1.0
Panel B: Y Variable values											
Indeed superior	0	0	0	0	0	0	0	0.02	0.18	0.98	1.0
Superior	0	0	0	0	0	0	0	0.10	0.30	0.90	1.0
More-or-less superior	0	0	0	0	0	0	0	0.32	0.55	0.95	1.0
Above average	0	0	0	0	0	0	0.20	0.65	0.90	0.95	1.0
Average	0	0.05	0.10	0.35	0.80	1.0	0.80	0.35	0.10	0.05	0
Below average	1.0	0.95	0.90	0.65	0.20	0	0	0	0	0	0
Poor	1.0	0.80	0.40	0.20	0.05	0	0	0	0	0	0
Very poor	1.0	0.64	0.16	0.04	0	0	0	0	0	0	0

Source: Adapted from Wilhelm and Parsaei (1991, pp. 194-195).

α = a compositional operator defined such that the elements of its membership function are determined as follows:

$$[f^{-1}{}_{M(x)}(\mu_x)\ \alpha\ f_{M(y)}(\mu_y)] = 1 \qquad \text{if } f^{-1}{}_{M(x)}(\mu_x) \leq f_{M(y)}(\mu_y)$$

or

$$[f^{-1}{}_{M(x)}(\mu_x)\ \alpha\ f_{M(y)}(\mu_y)] = f_{M(y)}(\mu_y) \qquad \text{if } f^{-1}{}_{M(x)}(\mu_x) > f_{M(y)}(\mu_y)$$

For example, if a company has the goal of 'getting high product quality' in its goal set, and this goal is judged to be '*very important*'; and if one of the available technologies is CAD and its capability to achieve this goal is considered to be '*more-or-less superior*', the r_{ij} (i refers to the goal and j refers to the technology) can be calculated:

- Using table (7.1) the membership functions of 'very important' for X and 'more-or-less superior' for Y are given as:

Very important (X) = [0 0.01 0.06 0.56 0.81 1.0 0.81 0.56 0.06 0.01 0]
More-or-less superior (Y) = [0 0 0 0 0 0 0 0.32 0.55 0.95 1.0]

- Using equation (7.11):

$$r_{ij} = \begin{bmatrix} 0 \\ 0.01 \\ 0.06 \\ 0.56 \\ 0.81 \\ 1.0 \\ 0.81 \\ 0.56 \\ 0.06 \\ 0.01 \\ 0 \end{bmatrix} (\alpha) \begin{bmatrix} 0 & 0 & 0 & 0 & 0 & 0 & 0 & 0.32 & 0.55 & 0.95 & 1.0 \end{bmatrix}$$

$$= \begin{bmatrix} 1.0 & 1.0 & 1.0 & 1.0 & 1.0 & 1.0 & 1.0 & 1.0 & 1.0 & 1.0 & 1.0 \\ 0 & 0 & 0 & 0 & 0 & 0 & 0 & 1.0 & 1.0 & 1.0 & 1.0 \\ 0 & 0 & 0 & 0 & 0 & 0 & 0 & 1.0 & 1.0 & 1.0 & 1.0 \\ 0 & 0 & 0 & 0 & 0 & 0 & 0 & 1.0 & 1.0 & 1.0 & 1.0 \\ 0 & 0 & 0 & 0 & 0 & 0 & 0 & 0.32 & 0.55 & 1.0 & 1.0 \\ 0 & 0 & 0 & 0 & 0 & 0 & 0 & 0.32 & 0.55 & 0.95 & 1.0 \\ 0 & 0 & 0 & 0 & 0 & 0 & 0 & 0.32 & 0.55 & 1.0 & 1.0 \\ 0 & 0 & 0 & 0 & 0 & 0 & 0 & 1.0 & 1.0 & 1.0 & 1.0 \\ 0 & 0 & 0 & 0 & 0 & 0 & 0 & 1.0 & 1.0 & 1.0 & 1.0 \\ 0 & 0 & 0 & 0 & 0 & 0 & 0 & 1.0 & 1.0 & 1.0 & 1.0 \\ 1.0 & 1.0 & 1.0 & 1.0 & 1.0 & 1.0 & 1.0 & 1.0 & 1.0 & 1.0 & 1.0 \end{bmatrix}$$

Similarly, for every combination of goals and available technologies a relation can be formulated. For example, if there are three goals and two available technologies then there will be six matrices. Each matrix that results from applying equation (7.11) represents an element in a new matrix (the columns represent the available technologies and the rows represent the goals).

$$f_{\Re j}(\mu_x, \mu_y) = \wedge f_{M(r)}(\mu_x, \mu_y) \qquad \text{over } i = 1, 2, \ldots\ldots, m \qquad (7.12)$$

The fourth step in the algorithm is to establish an 'ideal' value of X, denoted x^*, which represents the '*IMPORTANCE*' of implementing the overall AMT strategy, e.g., x^* = 'indeed critical'. The value of this variable is found in table (7.1). Then, for each available technology, a relational equation (7.13) can be solved.

$$Y_j^{-1} = \Re_j\,(\alpha)\, x^{*-1} \qquad \text{over } j = 1,2, \ldots\ldots,n \qquad (7.13)$$

where the membership function of Y_j^{-1} is determined by:

$$f^{-1}_{M(y)} = \wedge\, [f_{M(\Re)}(\mu_x, \mu_y)\ \alpha\ f^{-1}_{M(x^*)}(\mu_x)] \qquad (7.14)$$

This solution yields the membership function for the '*CAPABILITY*' of each of the available technologies (j =1,2, ...,n), over all m AMT goals, to implement the AMT strategy of the company. Wilhelm and Parsaei (1991) suggest that the analysis could stop here, and an attempt could then be made to assign linguistic meaning corresponding to the membership functions which result from the preceding computations, and the one(s) which are 'superior' could be selected. Alternatively, the analyst may wish to proceed to the final step.

In the final step, Wilhelm and Parsaei (1991) assume that the most desirable outcome of the decision process is the satisfaction of all AMT goals at the highest possible level by any technology selected. Thus, the membership functions resulting from the previous step should be compared with that corresponding to y^* = 'indeed superior'. This is done by computing the relative Hamming distance of each of the n membership functions resulting from the previous step from the 'ideal' using the following equation:

$$\delta_j = \frac{1}{s}\sum_{\mu_y}\left| f_{M(y)}(\mu_y) - f_{M(y^*)}(\mu_y)\right| \qquad \text{for } j = 1,2,\ldots\ldots,n \qquad (7.15)$$

where:

s : number of elements in the universe of discourse μ_y,
$f_{M(y)}(\mu_y)$: membership function for technology j from previous step,
$f_{M(y^*)}(\mu_y)$: membership function for technology y^*.

The best available technology is the one which has the minimum relative Hamming distance, i.e., the technology $J \in \{j = 1, 2, \ldots, n\}$ for which δ_j is a minimum.

The fuzzy linguistic approach described above provides a means of overcoming the main difficulty in evaluating AMT investment projects by permitting decision makers to express their judgements verbally. Accordingly, qualitative variables can be considered in the decision analysis. Also, the approach links the strategic goals of a firm and the capability of alternative technologies in achieving such goals. This link may force decision makers to take a long term view of their investment in AMT. However, as Wilhelm and Parsaei (1991, p. 202) point out 'this approach is still very much in the preliminary stages of development.' and three criticisms of the approach can be identified.

First, Wilhelm and Parsaei ignore an important part of AMT decision analysis by ignoring financial factors, or, at most treating them as if they are intangibles. For example, they include 'product cost' as one of the goals set and they use the linguistic variable '*CAPABILITY*' in representing this variable in the decision analysis. Second, Wilhelm and Parsaei concentrate on evaluating the ability of each technology to achieve the company's goals. However, the capability of a specific technology to achieve the company's goals does not mean the company can afford the investment required. Finally, in formulating the values of the linguistic variables, Wilhelm and Parsaei include two values with the same apparent meaning; 'unimportant' and 'not important', each of them having different membership functions. In addition, the membership functions of some variables seem counter-intuitive. For example, for the membership functions 'very important' and 'important', one might expect that points on the right side of the membership function of 'very important' would be higher than that of 'important'. However, Wilhelm and Parsaei's values are the reverse of this common sense expectation. As illustrated in figure 7.3, the membership function 'important' is greater than that for 'very important' on the right hand side.

In order to understand their apparently contradictory values, it is necessary to reflect on the meaning of the values. In panel A of table 7.1 there are three patterns. The first is from left to right on the universe of discourse, and contains the expressions '*unimportant*', '*important*', '*more-or-less critical*', '*critical*' and '*indeed critical*'. These are all matters of degree of *IMPORTANCE*. The second is the intensity of *IMPORTANCE*,

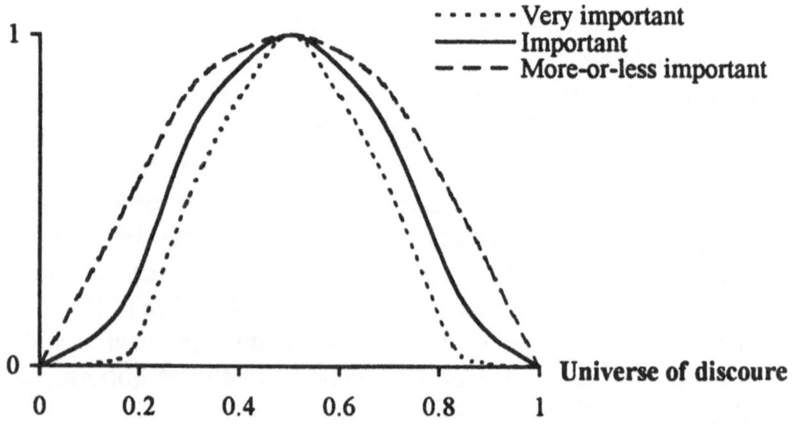

Figure 7.3 Membership functions of 'very important', 'important' and 'more-or-less important' values

which are '*more-or-less important*', '*important*', and '*very important*', which are illustrated in figure (7.3). Finally, there is the logical complement of '*important*', which is '*not important*', ('*important*' + '*not important*' = 1).

The first pattern is the intuitive use of the concept of *IMPORTANCE*. The latter two take the word '*important*' as the mid-point of the universe of discourse and use the word '*very*' and '*more-or-less*' as measures of intensity, i.e. fuzziness, of this mid-point; or the word '*not*' to mean not in the middle of the universe of discourse.

This illustrates one of the difficulties of using linguistic variables: that words and concepts are used differently by different people and in different context; and that precession may need to be defined during the process of evaluation.

A Proposed Model for AMT Investment Decisions

In the previous section two approaches to the use of fuzzy set theory in evaluating investment projects were reviewed. The first approach extends the net present value method to the case where the cash flows are represented by fuzzy numbers, while the second approach uses two linguistic variables (importance and capability) in evaluating AMT projects. In the following sections a model will be developed for evaluating AMT projects based on the conclusions of the empirical research (chapters 4 and 5) and utilising the concepts of fuzzy numbers and linguistic variables.

Figure (7.4), developed from figure (5.8), shows the framework of the suggested model. AMT investment decision making can be divided into three sequential stages: (1) examining the consistency of AMT investment projects with overall strategy, (2) measuring the expected performance of AMT investment projects, and (3) ranking AMT investment projects. Each stage is considered in the following sections.

Consistency with Overall Strategy

It was concluded from the field study that the consistency of AMT projects with overall strategy was an over-riding issue. Any AMT investment project should be consistent with the overall strategy before being subjected to detailed evaluation, and if an investment project is not consistent with strategy, it should be excluded from detailed evaluation. Thus, this stage in

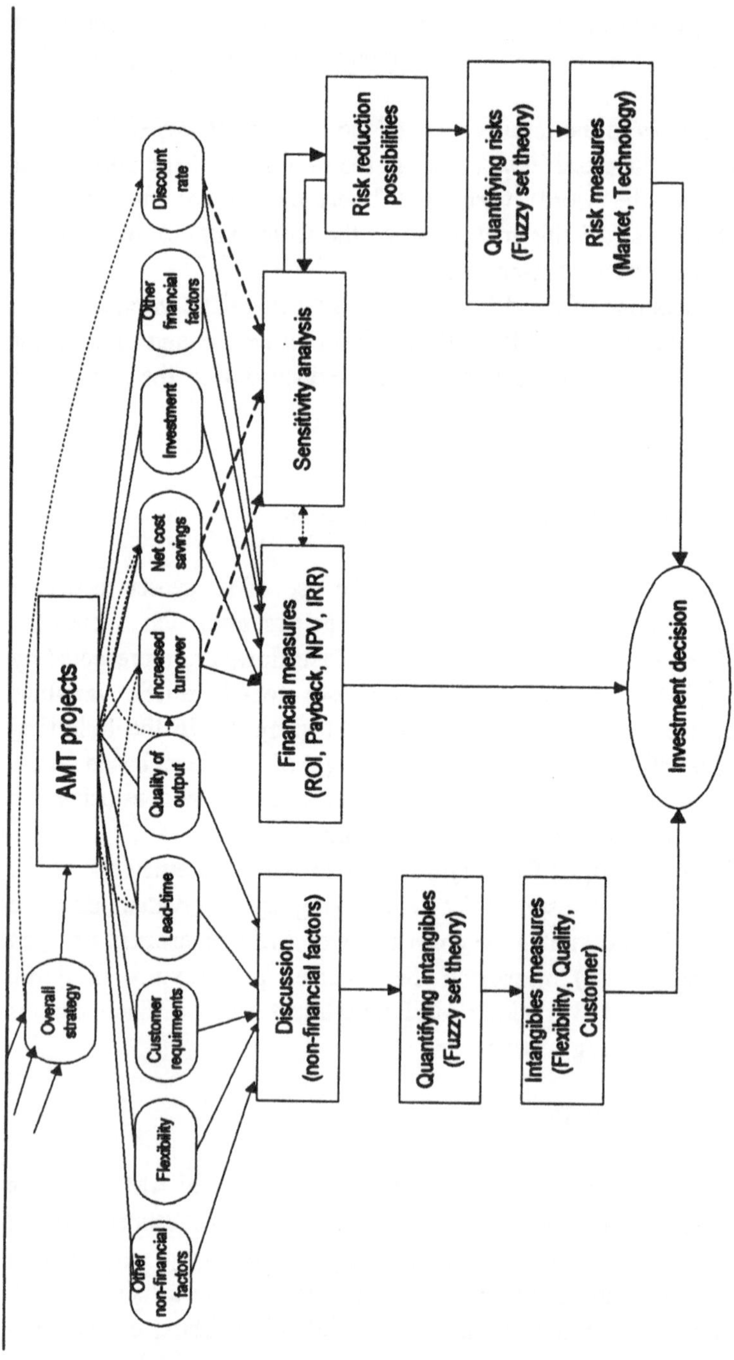

Figure 7.4 A model for AMT investment decision making

the evaluation of AMT investment projects is critical; only projects judged to be consistent with overall strategy should be considered for detailed evaluation.

There are various competitive strategies that a manufacturing company could adopt. Porter (1980) classified competitive strategies into two generic types: cost leadership and differentiation. Naik and Chakravarty (1992, p. 1585) criticised this classification as it: 'seems to be too broad and less useful for distinguishing manufacturing firms.' They identified seven general categories of competitive strategy that manufacturing firms adopt[3]. These are: 'innovation', 'customisation', 'product proliferation', 'price reduction', 'innovation and product proliferation', 'innovation and customisation', and 'product proliferation and price reduction'. Clearly the first four strategies are distinct from one another and the last three are combinations of the first four. Adopting one of these strategies may lead to specific types of manufacturing strategy with different options being considered for processing, material handling, tool handling, quality assurance and system integration (Naik and Chakravarty, 1992).

For example[4], adopting a 'product proliferation' strategy may be facilitated by acquiring: either multimachine flexible manufacturing lines or integrated CNC machines for the processing system; either an automated material handling system or an integrated material handling system with AS/RS; either automated tool delivery to machines or computer-controlled tool migration for the tool handling system, automated on-line inspection for quality assurance, and/or either integration with CAD/CAM or CIM.

Any AMT investment project should conform with the firm's strategy, and so this stage will limit the original alternatives to only those which satisfy the strategy. This leads to the second stage in which the expected performance of AMT investment projects is measured.

The Expected Performance of AMT Investment Projects

The expected performance of AMT investment projects can be based on the three measures of financial return, intangible benefits, and risk. Each measure is discussed as follows.

The Financial Return Measure The financial measure reflects all the factors that can be translated into cash flows. These factors include the investment amount, net cost savings, increased turnover and any other financial factors such as government grants and tax effects. Each factor is represented by a triangular fuzzy number described by three estimates: the lowest possible estimate, the best estimate, and the largest possible estimate. These three estimates are similar to the pessimistic, normal, and optimistic scenarios usually used in decision analysis. However, the proposed model allows decision makers to provide the required estimates as approximate values. For example, the investment amount could be specified as a value such as 'approximately £2 million' and this value may be interpreted as a triangular fuzzy number (1.9, 2, 2.1). If the value is an exact estimate, the triangular fuzzy number would be (2, 2, 2). The use of triangular fuzzy numbers makes the mathematics relatively tractable, and the results can be approximated to the triangular shape without losing significant information (Kaufmann and Gupta, 1991, pp. 55-67).

After estimating all the fuzzy cash flows for each alternative investment project, one or more financial measures: net present value, internal rate of return, payback and return on investment, can be used according to preference[5].

The Fuzzy Net Present Value (FNPV) Let *FNPV* be net present value, x_t be the net cash flow arising at the end of year *t*, *r* be the discount rate (assumed constant throughout the project life), I_0 be the initial investment at time *0*, and n be the project's life. Then *FNPV* is calculated as follows:

$$FNPV = \sum_{t=1}^{n} \frac{x_t}{(1+r)^t} - I_0 \tag{7.16}$$

As x_t, I_0, and r are triangular fuzzy numbers (x_{t1}, x_{t2}, x_{t3}), (I_{01}, I_{02}, I_{03}), *and* (r_1, r_2, r_3) respectively, the *FNPV* is also a triangular fuzzy number $(FNPV_1, FNPV_2, FNPV_3)$. Thus equation (7.16) is written as follows:

$$\begin{aligned}(FNPV_1, FNPV_2, FNPV_3) &= \sum_{t=1}^{n} \frac{(x_{t1}, x_{t2}, x_{t3})}{[(1,1,1)+(r_1, r_2, r_3)]^t} - (I_{01}, I_{02}, I_{03}) \\ &= \sum_{t=1}^{n} \frac{(x_{t1}, x_{t2}, x_{t3})}{[(1+r_1)^t, (1+r_2)^t, (1+r_3)^t]} - (I_{01}, I_{02}, I_{03}) \\ &= \sum_{t=1}^{n} \left(\frac{x_{t1}}{(1+r_3)^t}, \frac{x_{t2}}{(1+r_2)^t}, \frac{x_{t3}}{(1+r_1)^t} \right) - (I_{01}, I_{02}, I_{03}) \\ &= \left(\sum_{t=1}^{n} \frac{x_{t1}}{(1+r_3)^t} - I_{03}, \sum_{t=1}^{n} \frac{x_{t2}}{(1+r_2)^t} - I_{02}, \sum_{t=1}^{n} \frac{x_{t3}}{(1+r_1)^t} - I_{01} \right)\end{aligned} \tag{7.17}$$

where: $x_{t1} \geq o$

The membership function of *FNPV* is approximated to take the triangular shape as follows:

$$\mu_{FNPV}(y)=\begin{cases}(y-FNPV_1)/(FNPV_2-FNPV_1) & FNPV_1 \le y \le FNPV_2\\(FNPV_3-y)/(FNPV_3-FNPV_2) & FNPV_2 \le y \le FNPV_3\\0 & otherwise\end{cases} \quad (7.18)$$

It might be noted that the derivation of fuzzy present value as in 7.17 is rather similar to that of Chiu and Park (1994). However, here the three parameters of triangular fuzzy numbers are used for the fuzzy estimates instead of the left and right representations used in Chiu and Park's derivation. By using equation 7.17, decision analysts can calculate the fuzzy net present value in the usual way except that the calculation is repeated three times.

The Fuzzy Internal Rate of Return (FIRR) The internal rate of return is the rate of return which makes the *NPV* equal zero. To calculate the *IRR*, it is necessary to solve the following equation for r:

$$0=\sum_{t=1}^{n}\frac{x_t}{(1+r)^t}-I_0 \quad (7.19)$$

where r is the *IRR*. It is usually solved through a number of trials. Substituting each variable in the above equation with its triangular fuzzy number and using the same approach as in the *FNPV* equation 7.17, the fuzzy *IRR* (r_1, r_2, r_3) can be calculated using the following equation:

$$(0,0,0)=\left(\sum_{t=1}^{n}\frac{x_{t1}}{(1+r_3)^t}-I_{03},\sum_{t=1}^{n}\frac{x_{t2}}{(1+r_2)^t}-I_{02},\sum_{t=1}^{n}\frac{x_{t3}}{(1+r_1)^t}-I_{01}\right) \quad (7.20)$$

where $x_{t1} \geq o$ and (r_1, r_2, r_3) represents the *FIRR*, (IRR_1, IRR_2, IRR_3). The membership function of the *FIRR* is as follows:

$$\mu_{FIRR}(y)=\begin{cases}(y-FIRR_1)/(FIRR_2-FIRR_1) & FIRR_1 \le y \le FIRR_2\\(FIRR_3-y)/(FIRR_3-FIRR_2) & FIRR_2 \le y \le FIRR_3\\0 & otherwise\end{cases} \quad (7.21)$$

The Fuzzy Payback Period (FPB) The payback period is a measure of the time required to generate cash flows that recover the initial investment. If the initial investment or cash flows or both are fuzzy, the resultant payback period is also fuzzy. For example, consider a four-year project with estimated initial investment (I_0), in million, and cash flows (x_t), in million, (where t = 1,2,3,4) specified by triangular fuzzy numbers as I_0 = (£20, £25, £30), x_1 = (£7, £10, £12), x_2 = (£10, £12, £15), x_3 = (£10, £12, £15), x_4 = (£8, £10, £12). Assuming that x_t is realised at the end of year t, the fuzzy payback period (*FPB*), in years, is (2, 3, 4).

It should be noted that the first parameter in I_0 [£20] is recovered from the third parameters in x_t [£12, £15, £15, £12], and so the initial investment of £20 million cannot be recovered before the end of the second year, the second parameter in I_0 [£25] is recovered from the second parameters in x_t [£10, £12, £12, £10], and the third parameter in I_0 [£30] is recovered from the first parameters in x_t [£7, £10, £10, £8]. This calculation uses the rules for subtraction of triangular fuzzy numbers.

The Fuzzy Return on Investment (FROI) ROI can be calculated in several different ways (Lumby, 1995). One way is to divide the average accounting profit generated over the life of the project by the initial outlay. The annual accounting profit can be calculated by deducting depreciation from the sum of net cash flows. Then, the average accounting profit over the project's life is the total of annual accounting profits divided by the length of the project's life. If either the accounting profit or initial outlay or both are expressed as fuzzy numbers, then the ROI is also a fuzzy number.

Let *FROI* [($FROI_1$, $FROI_2$, $FROI_3$)] be the fuzzy return on investment, I_0 [(I_{01}, I_{02}, I_{03})] be the fuzzy initial outlay, D be the total depreciation over the project's life, x_t [(x_{t1}, x_{t2}, x_{t3})] be fuzzy net cash flows for the year t (t = $1,2,\dots,n$), and s [(s_1, s_2, s_3)] be the fuzzy scrap value at end of last year of the project. Based on the following assumptions: (1) the initial outlay minus scrap value is depreciable over the project's life, (2) depreciation is the only non-cash expense over the project's life, and (3) the straight-line method is used in calculating depreciation, The fuzzy *FROI* is calculated as follows:

$$FROI = \frac{(\sum_{t=1}^{n} x_t - D)/n}{I_0} \tag{7.22}$$

Then, substituting each variable by its triangular fuzzy number, the *FROI* is:

$$(FROI_1, FROI_2, FROI_3) = \frac{(\sum_{t=1}^{n}(x_{t1}, x_{t2}, x_{t3}) - [(I_{01}, I_{02}, I_{03}) - (s_1, s_2, s_3)]) / n}{(I_{01}, I_{02}, I_{03})} \tag{7.23}$$
$$= \left(\frac{1}{nI_{03}}(\sum_{t=1}^{n} x_{t1} - I_{03} + s_1), \frac{1}{nI_{02}}(\sum_{t=1}^{n} x_{t2} - I_{02} + s_2), \frac{1}{nI_{01}}(\sum_{t=1}^{n} x_{t3} - I_{01} + s_3)\right)$$

The membership function of the *FROI* is as follows:

$$\mu_{FROI}(y) = \begin{cases} (y - FROI_1)/(FROI_2 - FROI_1) & FROI_1 \le y \le FROI_2 \\ (FROI_3 - y)/(FROI_3 - FROI_2) & FROI_2 \le y \le FROI_3 \\ 0 & otherwise \end{cases} \tag{7.24}$$

For example, if the scrap value s = (£3, £4, £5) with the same I_o, x_t as provided in the previous numerical example, the *FROI* is calculated as follows:

$$FROI_1 = \frac{1}{nI_{03}}(\sum_{t=1}^{4} x_{t1} - I_{03} + s_1),$$
$$= \frac{1}{(4)(30)}(7 + 10 + 10 + 8 - 30 + 3) = 0.07$$
$$FROI_2 = \frac{1}{nI_{02}}(\sum_{t=1}^{4} x_{t2} - I_{02} + s_2),$$
$$= \frac{1}{(4)(25)}(10 + 12 + 12 + 10 - 25 + 4) = 0.23$$
$$FROI_3 = \frac{1}{nI_{01}}(\sum_{t=1}^{4} x_{t3} - I_{01} + s_3),$$
$$= \frac{1}{(4)(20)}(12 + 15 + 15 + 12 - 20 + 5) = 0.49$$

$\therefore$ *ROI* = (7%, 23%, 49%)

Representation of the Financial Return Measure ($FRMA_K$) in the model
Choosing financial measure(s) to evaluate alternative investment projects

depends mainly on the preference of the decision maker(s). Most texts recommend one of the discounted cash flow methods, and typically, prefer NPV as it is consistent with the objective of maximising share-holders' wealth. However, most practitioners consider non-discounted measures (payback period and ROI) to be more important than their counterparts. Further, when they use a discounting method, IRR is usually preferred[6].

In the model suggested here, the decision maker is not limited to a specific financial return measure, and computing more than one financial return measure may provide decision makers with more information. In principle, it would be possible to combine two or more financial measures into a single project score[7]. However, it is recommended here that only one financial return measure (either *FNPV* or *FIRR*) be used to represent the project's profitability. This is because the same data is used to produce all the financial return measures, and thus, representing the financial return measure in the model by more than one parameter may lead to duplication.

The Non-financial Measure The non-financial measure reflects all of the factors that cannot be translated into cash flows. In order to generate a measure for this group of factors, a hierarchical analysis with three levels is used. The first two levels evaluate the fuzzy importance of the non-financial investment criteria, and sub-criteria and the lowest level permits ratings to be assigned to alternative AMT investment projects. The hierarchical structure is shown in figure 7.5.

The hierarchy shown in figure 7.5 is based on the empirical research. The non-financial criteria included in the model are these identified by practitioners as important in AMT evaluation. The non-financial criteria include 'improving product quality', 'increasing process flexibility' and 'meeting customer requirements'[8]. In order to avoid confusion between these criteria, 'product quality' is defined as *conforming with product specification* while 'customer requirements' is related to *development of new product features*. Process flexibility might be usefully divided into two sub-dimensions: relating to *handling increased product complexity* and *increased product volume*.

Some intangible benefits cannot be readily integrated into the model, in particular the closely related benefits of inventory and lead-time reduction. The reduction of inventory is seen, in simplistic JIT thinking, as a good thing in itself. However, closer consideration suggests that the advantages of reduced inventory can be analysed as a number of separate factors: reduced lead-time, greater responsiveness to customers, lower costs of quality

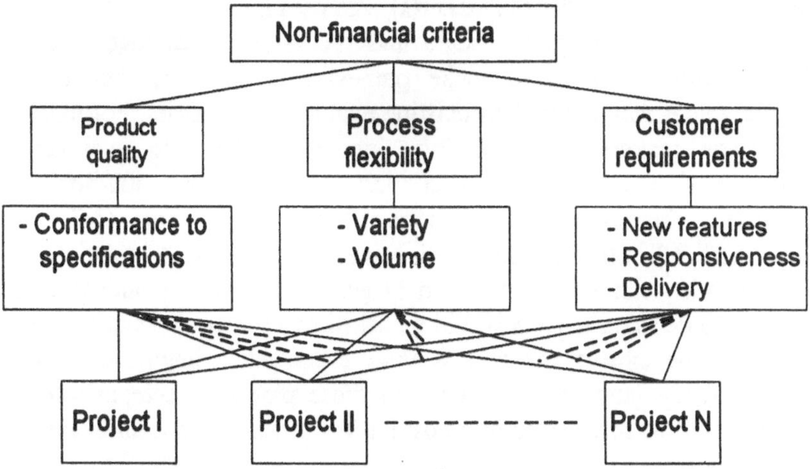

Figure 7.5 The non-financial criteria hierarchy

(because defects and errors are identified more rapidly), better production scheduling (because there are no buffer stocks), and reduced inventory holding cost. At first sight, it seems that reduced lead-times and inventory are process related benefits, but it is more logical to include the specific benefits in the appropriate part of the decision model. Thus, financial savings from reduced inventory and quality costs can be incorporated into the calculation of financial return; reduced production lead-time and better response to customers can be considered as part(s) of customer requirements, improved quality can be considered under product quality.

Developing a Fuzzy Linguistic Scale A fuzzy linguistic scale is proposed to evaluate the fuzzy importance of the non-financial investment criteria and sub-criteria in the first two levels of the hierarchy [figure 7.5], and to assign fuzzy ratings to the alternative AMT investment projects in the lowest level of the hierarchy. Previous studies have usually suggested generalised linguistic variables, but here the linguistic variable concept is used in a rather different way. The linguistic scale is generated *by decision makers themselves* because the meaning (values) of any word (linguistic variable) depends mainly on the decision maker(s) who uses it.

To generate a linguistic scale and its values (meanings), decision makers would be asked to make pair-wise comparisons between the linguistic values they use in their existing evaluation processes. For example, the decision makers may use five linguistic values, to evaluate the importance of non-financial criteria: very important (*VI*), important (*I*), more-or-less important (*MI*), more-or-less unimportant (*MU*) and unimportant (*U*). To elicit the relative meaning of these linguistic values (their equivalent numbers to be used in generating a non-financial criteria measure) the decision makers would be asked to express by what multiple a linguistic value exceeds or is less than every other linguistic value. If triangular fuzzy numbers were used in these pair-wise comparisons, the answers would be formulated in a positive reciprocal matrix, an example of which follows:

$$\begin{array}{c} \\ VI \\ I \\ MI \\ MU \\ U \end{array} \begin{array}{c} \begin{array}{ccccc} VI & I & MI & MU & U \end{array} \\ \begin{bmatrix} (1,1,1) & (2,3,4) & (4,5,6) & (6,7,8) & (8,9,9) \\ (1/4,1/3,1/2) & (1,1,1) & (1,2,3) & (2,3,4) & (3,4,5) \\ (1/6,1/5,1/4) & (1/3,1/2,1) & (1,1,1) & (1,3/2,2) & (3/2,2,3) \\ (1/8,1/7,1/6) & (1/4,1/3,1/2) & (1/2,2/3,1) & (1,1,1) & (1,5/4,3/2) \\ (1/9,1/9,1/8) & (1/5,1/4,1/3) & (1/3,1/2,2/3) & (2/3,4/5,1) & (1,1,1) \end{bmatrix} \end{array}$$

To compute the fuzzy rate of each linguistic value, the geometric mean method is used. This method is chosen for its simplicity and its suitability in fuzzy positive reciprocal matrices (Buckley, 1985)[9]. Given the positive comparison matrix as:

$$A = \begin{bmatrix} a_{11} & a_{12} & \cdots & a_{1n} \\ a_{21} & a_{22} & & a_{2n} \\ \vdots & & & \\ a_{n1} & a_{n2} & & a_{nn} \end{bmatrix} \tag{7.25}$$

The geometric mean of each row is calculated as:

$$Q_i = \left[\prod_{j=1}^{n} a_{ij} \right]^{1/n} \tag{7.26}$$

The weight W_i is then calculated for each row as:

$$W_i = Q_i / (\sum_{i=1}^{n} Q_i) \tag{7.27}$$

Let a_{ij} be a triangular fuzzy number (b_{ij}, c_{ij}, d_{ij}), Q_i be a triangular fuzzy number (e_i, f_i, g_i) and W_i be a triangular fuzzy number (k_i, l_i, m_i), then the previous two equations can be written as:

$$Q_i = (e_i, f_i, g_i) = \left(\left[\prod_{j=1}^{n} b_{ij} \right]^{1/n}, \left[\prod_{j=1}^{n} c_{ij} \right]^{1/n}, \left[\prod_{j=1}^{n} d_{ij} \right]^{1/n} \right) \tag{7.28}$$

$$W_i = (k_i, l_i, m_i) = \left(\left[e_i / \sum_{i=1}^{n} g_i \right], \left[f_i / \sum_{i=1}^{n} f_i \right], \left[g_i / \sum_{i=1}^{n} e_i \right] \right) \tag{7.29}$$

The triangular fuzzy number of each linguistic value can then be calculated as the accumulated sum of this linguistic value.

To illustrate the procedures to derive a fuzzy linguistic scale, the previous pair-wise comparison matrix is used. Table 7.2 gives the results of calculating the geometric mean (Q_i) [equation 7.28] for each row in the pair-wise comparisons while table 7.3 gives the weights (W_i) calculated for each linguistic value.

In principle, the weights (W_i) computed in table 7.3 could be used as fuzzy meanings for their equivalent linguistic values in developing measures for non-financial criteria. However, it is preferred to transfer these weights into a rating scale. This can be done for each linguistic value by adding all lowest weights to the weight of this linguistic value. Using the results provided in table 7.3. The rating scale would be as follows:

U = (0.04, 0.06, 0.08),
MU = (0.04, 0.06, 0.08) (+) (0.05, 0.07, 0.11) = (0.09, 0.13, 0.19),

Table 7.2 The geometric mean for the linguistic values

Q_i	e_i		f_i		g_i	
Q_1 (*VI*)	$(1\times2\times4\times6\times8)^{1/5}$	= 3.29	$(1\times3\times5\times7\times9)^{1/5}$	= 3.94	$(1\times4\times6\times7\times9)^{1/5}$	= 4.32
Q_2 (*I*)	$(1/4\times1\times2\times2\times3)^{1/5}$	= 1.24	$(1/3\times1\times2\times3\times4)^{1/5}$	= 1.51	$(1/2\times1\times3\times4\times5)^{1/5}$	= 1.97
Q_3 (*MI*)	$(1/6\times1/3\times1\times1\times3/2)^{1/5}$	= 0.61	$(1/5\times1/2\times1\times3/2\times2)^{1/5}$	= 0.79	$(1/4\times1\times1\times2\times3)^{1/5}$	= 1.08
Q_4 (*MU*)	$(1/8\times1/4\times1/2\times1\times1)^{1/5}$	= 0.43	$(1/7\times1/3\times2/3\times1\times5/4)^{1/5}$	= 0.52	$(1/6\times1/2\times1\times1\times3/2)^{1/5}$	= 0.66
Q_5 (*U*)	$(1/9\times1/5\times1/3\times2/3\times1)^{1/5}$	= 0.34	$(1/9\times1/4\times1/2\times4/5\times1)^{1/5}$	= 0.41	$(1/8\times1/3\times2/3\times4/5\times1)^{1/5}$	= 0.47
Total		5.91		7.17		8.50

Table 7.3 Weights of the linguistic values

W_i	$k_i = e_i / 8.50$	$l_i = f_i / 7.17$	$m_i = g_i / 5.91$
W_1 (*VI*)	0.39	0.55	0.73
W_2 (*I*)	0.14	0.21	0.33
W_3 (*MI*)	0.07	0.11	0.18
W_4 (*MU*)	0.05	0.07	0.11
W_5 (*U*)	0.04	0.06	0.08

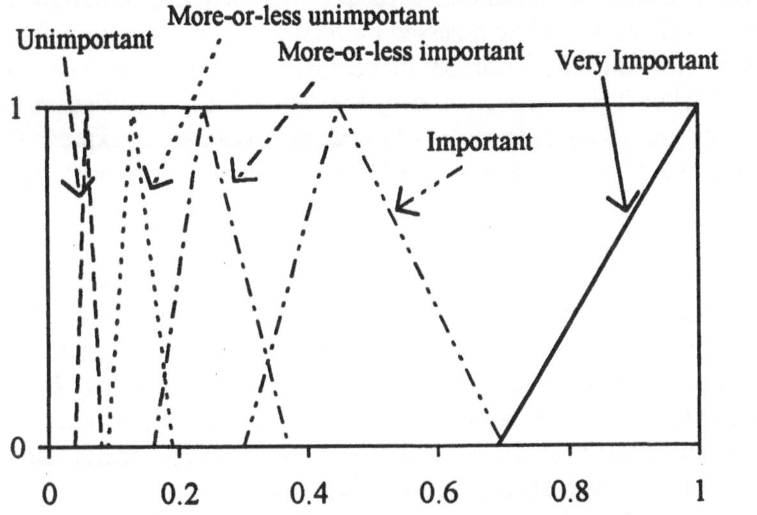

Figure 7.6 Membership functions of linguistic scale

MI = (0.09, 0.13, 0.19) (+) (0.07, 0.11, 0.18) = (0.16, 0.24, 0.37),
I = (0.16, 0.24, 0.37) (+) (0.14, 0.21, 0.33) = (0.30, 0.45, 0.70),
VI = (0.30, 0.45, 0.70) (+) (0.39, 0.55, 0.73) = (0.69, 1.00, 1.00[10])

Accordingly, the fuzzy linguistic scale (*LS*) is:

LS = {U, MU, MI, I, VI} = {(0.04, 0.06, 0.08), (0.09, 0.13, .19), (0.16, 0.24, .37), (0.30, 0.45, 0.70), (0.69, 1.00, 1.00)}

The membership functions of each linguistic value in *LS* are shown in figure 7.6.

If there were several decision makers, each decision maker would be asked separately for his or her pair-wise comparisons. All of the decision makers' matrices would then be combined by taking the arithmetic mean of their values. Then, the fuzzy linguistic rating scales would be calculated in the same way as above.

Developing a Fuzzy Non-financial Measure The weighted importance of non-financial criteria would be assessed by requesting the decision maker to weight each non-financial criterion in comparison with each other non-financial criterion. This importance weight is denoted (C_i). Decision makers would use the fuzzy linguistic scale (*LS*) already developed, to express their views. This would allow them to make a subjective assessment of the importance of the different non-financial criteria without the pressure of being precise (Naik and Chakravarty, 1992).

The decision maker would then be asked to rate each alternative with respect to each non-financial criterion using a fuzzy linguistic scale. Here, another linguistic scale may be developed (in the same way as the one developed earlier for the importance of non-financial criteria) to suit the terms used in rating an alternative's ability. Table 7.4 presents two linguistic scales that can be used for the importance of non-financial criteria and for the alternatives' ability to achieve sub-criterion requirements.

Suppose A_k denotes one alternative project k (k = 1, 2, ... , m alternatives), and LSA_{ki} denotes the fuzzy linguistic rating for alternative A_k with respect to a non-financial criterion i (i = 1, 2, ... , n non-financial criteria).

The fuzzy non-financial measure of alternative A_k, denoted $FNFMA_k$, can then be computed as follows:

$$FNFMA_k = \frac{1}{n}\left(\sum_{i=1}^{n}(C_i)(LSA_{ki})\right) \tag{7.30}$$

Table 7.4 Linguistic scales for non-financial criteria

Importance of non-financial criteria/sub-criteria		Alternative's ability to achieve sub-criterion requirements	
(.04, .06, .08)	Unimportant	(.00, .10, .19)	Very low
(.09, .11, .19)	More-or-less unimportant	(.15, .25, .40)	Low
(.16, .24, .37)	More-or-less important	(.30, .50, .70)	Medium
(.30, .45, .70)	Important	(.65, .75, .90)	High
(.69, 1.00, 1.00)	Very important	(.70, 1.00, 1.00)	Very high

Let C_i be a triangular fuzzy number (V_{i1}, V_{i2}, V_{i3}) and LSA_{ki} be a triangular fuzzy number (G_{ki}, H_{ki}, I_{ki}), then $FNFMA_k$ is also a triangular fuzzy number (W_{k1}, W_{k2}, W_{k3}), and so equation 7.30 can be written as:

$$FNFMA_k = (W_{k1}, W_{k2}, W_{k3}) = \left(\left[\frac{1}{n}\left(\sum_{i=1}^{n}(V_{i1})(G_{ki})\right)\right], \left[\frac{1}{n}\left(\sum_{i=1}^{n}(V_{i2})(H_{ki})\right)\right], \left[\frac{1}{n}\left(\sum_{i=1}^{n}(V_{i3})(I_{ki})\right)\right]\right) \quad (7.31)$$

The membership function of the fuzzy non-financial criteria measure ($FNFMA_k$) is as follows:

$$\mu_{FNFMA_k}(y) = \begin{cases} (y - W_{k1})/(W_{k2} - W_{k1}) & W_{k1} \le y \le W_{k2} \\ (W_{k3} - y)/(W_{k3} - W_{k2}) & W_{k2} \le y \le W_{k3} \\ 0 & otherwise \end{cases} \quad (7.32)$$

The Risk Measure The empirical study concluded that most practitioners in the companies sampled combine both moment-oriented and dimension-oriented approaches in dealing with investment project risk. The first step in analysing project risk is to conduct sensitivity analysis in order to identify the most sensitive variables. Possibilities of reducing project risk are then identified and every effort made to minimise or eliminate any obvious risk regarding these variables. Finally, the residual risk, which cannot be minimised or eliminated, can be reflected by using high discount rates and a shorter payback period. However, none of the interviewees suggested a formal basis to make such adjustments to the discount rate or target payback period. Also, many interviewees indicated that they did not change the discount rate according to the level of the risk inherent in each investment project, often industry averages were used for all projects regardless of the expected level of project risk.

Adjusting discount rates for risk has been criticised by many authors (for example: Ronen and Sorter, 1972; Kaplan and Atkinson, 1989; Accola, 1994; and Accola, et al., 1995) because it is difficult to capture all the aspects of project risk through a single modification of the discount rate. In addition, adjustments to the discount rate are affected by managers' *attitudes* toward risk rather than by an explicit representation of the individual risks inherent in each of the investment alternatives[11]. Accola (1994, p. 21) argued that:

> '... Models used should enable managers to measure and evaluate expected cash flows separately from the risks associated with them. In choice situations involving multiple aspects of risk, models should be used to explicitly represent all significant aspects of risk existing in the choice alternatives'.

In the model proposed here, instead of adjusting the discount rate for risk, risk is represented by a separate dimension. Therefore, decision makers are provided with information that reflects the level of risk inherent in each alternative investment project. Two types of AMT investment project risk were discussed in the field study: risk relating to market and risk relating to technology. Each type of risk can thus be represented by a sub-dimension. A hierarchical representation of project risks is shown in figure 7.7.

The decision maker would be asked to weight each risk using a fuzzy linguistic scale. The weight assigned to a particular risk factor is denoted (WR_j) (where j =1, 2, ...m risk factors under each type of risk).

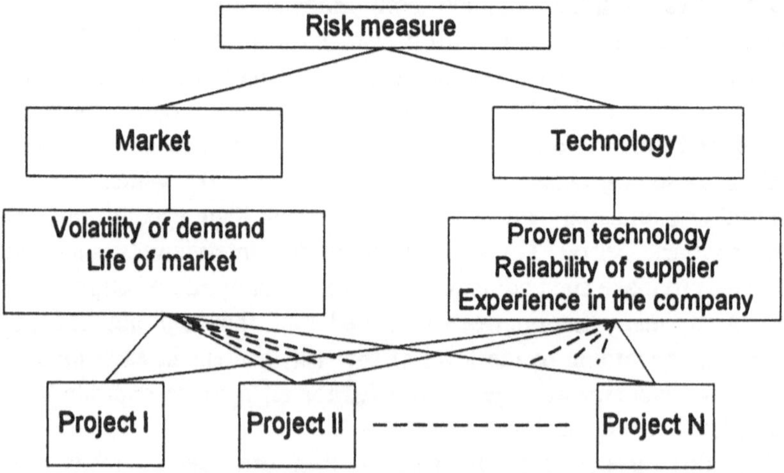

Figure 7.7 The risk measure hierarchy

The decision maker would then evaluate each alternative AMT project with respect to each risk factor. This evaluation could be done qualitatively (using the linguistic scale) or quantitatively (using measures as proxies). For example, proven technology could be expressed qualitatively as very high or quantitatively by how many times this type of technology has been applied in the industry and/or how long the type of technology has been in use. Reliability of supplier could be represented by the size of the supplier measured by, for example, turnover or total assets. Experience in the company could be represented by the number of occasions that similar technology has been applied at both plant level and company wide. If the decision maker prefers to evaluate the riskiness of alternative projects with respect to any risk factor quantitatively, this evaluation should first be converted into a score in the closed interval [0,1] by normalising its values. Suppose a decision maker evaluates a project A_k (k = 1, 2, ... N projects) with respect to a risk factor R_j (j = 1, 2, ... m risk factors) as $\dot{RA}_{kj}$, the converted value is denoted as RA_{kj} and computed as follows:

$$RA_{kj} = (\dot{RA}_{kj} \,/\, \sum_{i=1}^{k} \dot{RA}_{kj}) \,/\, a \tag{7.33}$$

where: a is a suitable non-fuzzy constant so that all supports of RA_{kj} are in the closed interval [0,1].

If the decision maker prefers to evaluate the riskiness of alternative projects qualitatively using a linguistic rating scale, the corresponding fuzzy number is used in the subsequent procedures. RA_{kj} is a fuzzy number denoting the riskiness of a project A_k with respect to risk factor j whether it is a converted value or the corresponding value on the linguistic scale.

It should be noted that not all risk factors indicate the same preference. For example, a linguistic value of 'very high' for 'volatility of demand' means higher risk than a linguistic value of 'high', while 'very high' for 'reliability of supplier' means the opposite. Thus, for consistency, the corresponding fuzzy numbers in the fuzzy linguistic scale must take this into account. Another fuzzy linguistic scale can be developed in the same way as stated earlier. Example fuzzy numbers associated with the fuzzy linguistic scale for each risk factor are presented in table 7.5.

Table 7.5 The fuzzy linguistic scale for risk factors

Fuzzy numbers	VD	LM	PT	RS	EC
(0.00, 0.06, 0.08)	Non	Infinite	Very high	Very high	Excellent
(0.09, 0.11, 0.19)	Very low	Very long	High	High	Very good
(0.16, 0.24, 0.37)	Low	Long	Average	Average	Good
(0.30, 0.45, 0.70)	Average	Average	Low	Low	Average
(o.60, 0.70, 0.85)	High	short	Very low	Very low	Poor
(0.80, 1.00, 1.00)	Very high	Very short	Non	Non	Very poor

VD: Volatility of demand, LM: Life of market, PT: Proven technology, RS: Reliability of supplier, EC: Experience in the company.

The fuzzy risk measure, represented by $FRMA_k$, can then be calculated as follows:

$$FRMA_k = \frac{1}{m}\left(\sum_{j=1}^{m} (WR_j)(RA_{jk})\right) \tag{7.34}$$

Let WR_j and RA_{jk} be triangular fuzzy numbers (WR_{j1}, WR_{j2}, Wr_{j3}) and (RA_{jk1}, RA_{jk2}, RA_{jk3}) respectively, then $FRMA_k$ will also be a triangular fuzzy number ($FRMA_{k1}$, $FRMA_{k2}$, $FRMA_{k3}$). Hence, equation 7.34 can be written as follows:

$$\begin{aligned} FRMA_k &= (FRMA_{k1}, FRMA_{k2}, FRMA_{k3}) \\ &= \left(\left[\frac{1}{m}\left(\sum_{j=1}^{m}(WR_{j1})(RA_{jk1})\right)\right], \left[\frac{1}{m}\left(\sum_{j=1}^{m}(WR_{j2})(RA_{jk2})\right)\right], \left[\frac{1}{m}\left(\sum_{j=1}^{m}(WR_{j3})(RA_{jk3})\right)\right]\right) \end{aligned} \tag{7.35}$$

The membership function of the fuzzy risk measure ($FRMA_k$) is as follows:

$$\mu_{FRMA_k}(y) = \begin{cases} (y - FRMA_{k1})/(FRMA_{k2} - FRMA_{k1}) & FRMA_{k1} \le y \le FRMA_{k2} \\ (FRMA_{k3} - y)/(FRMA_{k3} - FRMA_{k2}) & FRMA_{k2} \le y \le FRMA_{k3} \\ 0 & Otherwise \end{cases} \tag{7.36}$$

Ranking AMT Investment Projects

After developing the three fuzzy measures of financial return, non-financial criteria and risk for each AMT project under consideration, the decision maker has to choose the 'best' AMT project. To do this each fuzzy measure needs to be reduced to an equivalent crisp measure so that relative differences between AMT projects for each measure can be identified, and a rule for selecting the 'best' AMT project is needed.

Reducing Fuzzy Measures into Crisp Measures Each fuzzy measure is represented by a triangular fuzzy number, and so the decision maker would have three values for each measure computed for each AMT project. Reducing these three values into an equivalent crisp value can be achieved by ranking the fuzzy numbers. A number of ranking methods were reviewed and a further method was proposed in the previous chapter. Using the proposed ranking method, the crisp or reduced value for each measure can be calculated as follows.

Let FMA_k be a triangular fuzzy measure (FMA_{k1}, FMA_{k2}, FMA_{k3}) for project A_k (k =1, 2, … , N), the reduced value for project A_k (RVA_k) can be computed as:

$$\begin{aligned} RV(A_k) = (FMA_{k2})\{(O)[(FMA_{k3} - x_{\min}) / (x_{\max} - x_{\min} + FMA_{k3} - FMA_{k2})] \\ +(1-O)[1-(x_{\max} - FMA_{k1}) / (x_{\max} - x_{\min} + FMA_{k2} - FMA_{k1})]\} \end{aligned} \tag{7.37}$$

where:

$$x_{min} = inf\, S \tag{7.38}$$

$$x_{max} = sup\, S \tag{7.39}$$

$$S = \bigcup_{k=1}^{n} S_k \tag{7.40}$$

$$S_k = \left\{FMA_{k1}, FMA_{k2}, FMA_{k3} \middle| \forall\, k = 1,2,\cdots N\right\} \tag{7.41}$$

and:

O : an index of optimism in the closed interval [0,1]

Selecting the 'Best' AMT Project Using equation 7.37 for each of the three fuzzy measures will produce three reduced values for each AMT project. The fuzzy measures of financial return and non-financial criteria were formulated as 'higher value is better', and so the reduced value retains the same feature that higher values for financial return or for non-financial criteria measure will be preferred. For the risk measure however, the lower value is better as it means less associated risk. In principle it would be possible to combine the three major dimensions of financial return, non-financial criteria and risk into a single project score. However, this combination would mean a significant loss of information. Instead, using the 'efficient plane' suggested by Accola (1994), decision makers are provided with more information regarding each dimension for the AMT investment projects being evaluated. The efficient plane is represented by all alternative AMT investment projects that dominate other alternatives below the plane. In figure 7.8, the shaded area represents an efficient plane which is

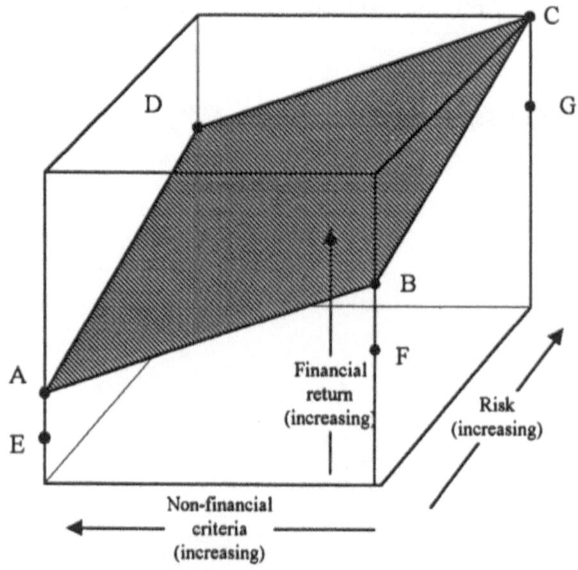

Figure 7.8 A three dimensional analysis for selecting AMT investment projects

Source: Adapted from Accola (1994, p. 24).

very risky and provides highest financial return. Alternative D provides non-financial benefits as much as alternative A, but it has higher risk and higher financial return. The selection of an alternative from the set of alternatives on the efficient plane depends on the relative preferences of decision maker(s) for risk, non-financial benefits and financial return.

The efficient plane slopes downward as non-financial benefits increase, based on the assumption that the financial return for investment projects with large non-financial benefits is likely to be less than the financial return for investment projects with few non-financial benefits. However, the slope of the efficient plane will vary, depending on the set of alternatives under consideration by a company (Accola, 1994).

An algorithm which summarises the proposed model for AMT investment decisions is provided in table 7.6.

An Illustrative Example

In this section, a hypothetical AMT investment decision is outlined to demonstrate the computational process of the proposed AMT investment decision model summarised in table 7.6.

The example involves a large engineering company that produces a number of products sold mainly to other industrial companies in the UK and overseas. The current manufacturing system is based on an operator-controlled NC machine cell. In order to maintain its competitive position, the company is considering an investment in an advanced manufacturing system. After an initial screening, three alternative AMT projects I, II and III in addition to the current manufacturing system, are to be evaluated. Project I consists of multi-machine CNC, operator-assisted MHS, manual tool changeover and manual off-line inspection. Project II consists of integrated CNC machine, computer controlled MHS, automated tool delivery and automatic on-line inspection. Project III is a fully integrated manufacturing system which consists of integrated CNC machine, integrated MHS with AS/RS, computer controlled tool migration, feedback for automatic process control and CAD/CAM. The steps in the suggested procedure are as follows.

Table 7.6 Stepwise description of the AMT investment decision algorithm

1. Form a committee of decision makers, decide the appropriate financial measure, non-financial criteria and risk measures, and identify the available alternative AMT investment projects.
2. Check each alternative for consistency with overall company strategy. Consider only those alternatives which have the desired consistency.
3. Develop fuzzy estimates of project cash flows, including those items which can be sensibly quantified in cash flow terms, and calculate the appropriate fuzzy financial measure(s).
4. Perform sensitivity analysis in order to identify any key variables which might be especially sensitive ('risky').
5. Identify intangible benefits that have not been quantified in the financial analysis. These benefits will probably fall under three headings: product, process and market. A hierarchical structure may be useful in summarising the attributes of each alternative.
6. Either adopt an existing fuzzy linguistic scale or develop one using pair-wise comparisons to evaluate the fuzzy importance of the non-financial investment criteria and sub-criteria, and to assign fuzzy ratings to the AMT investment projects.
7. Tabulate the importance assigned to each of the non-financial criteria and sub-criteria by the decision maker(s).
8. Tabulate the ratings assigned to each alternative with respect to each sub-criterion by the decision maker(s).
9. Calculate the fuzzy non-financial criteria measure for each alternative using equation 7.31.
10. Consider any action which might reduce the risk inherent in sensitive variables. Then identify the residual risk associated with each alternative. A hierarchical representation may be useful in summarising inherent risks and these will probably fall under two headings: market and technology.
11. Weight each risk factor using a fuzzy linguistic scale (or quantitative proxies) and evaluate each alternative with respect to each risk factor.
12. Calculate the fuzzy risk measure for each alternative ($FRMA_k$) using equation 7.35.
13. Calculate the ranked value for each alternative under each fuzzy measure (financial, non-financial and risk) using equations 7.37 - 7.41.
14. Develop an efficient plane using ranked values calculated in the previous step and choose an alternative located on this plane which satisfies the decision maker's preferences regarding financial return, non-financial benefits and risk.

Step 1

A committee of managers of the finance, production and marketing departments is formed to evaluate the available alternatives. Four alternatives are considered: the current manufacturing system (CS), project I, project II and project III. The company uses NPV and payback period as financial return measures. The non-financial criteria to be employed include 'product quality', 'process flexibility' and 'customer requirements'. Both market and technology risks will be measured.

Step 2

The company adopts an overall strategy of 'innovation and product proliferation'. The committee believes that the CS is not consistent with this strategy. Hence, the CS alternative is omitted and only the other three alternatives are considered.

Step 3

It is assumed that the planning horizon (n) is 7 years and the fuzzy discount rate (r) is (7%, 8%, 10%). The initial investments (I_o) and net cash flows after tax (x_t) at end of year t to the planned horizon are as in table 7.7.

Using equation 7.17 the FNPV for project I can be calculated as follows:

Table 7.7 Estimated fuzzy initial investments (I_0) and fuzzy net cash flows after tax (x_t) (£000,000s)

	Project I	Project II	Project III
I_o	(0.8, 1.0, 1.1)	(1.2, 1.4, 1.6)	(1.6, 1.8, 1.9)
x_1	(.33, .35, .37)	(.38, .40, .41)	(.41, .42, .43)
x_2	(.33, .35, .37)	(.38, .40, .41)	(.41, .42, .43)
x_3	(.33, .35, .37)	(.38, .40, .41)	(.41, .42, .43)
x_4	(.33, .35, .37)	(.38, .40, .41)	(.41, .42, .43)
x_5	(.33, .35, .37)	(.38, .40, .41)	(.41, .42, .43)
x_6	(.31, .33, .35)	(.37, .39, .41)	(.41, .42, .43)
x_7	(.28, .30, .31)	(.34, .36, .40)	(.38, .40, .42)

$$FNPV_1 = \left(\left[\frac{.33}{(1+.10)^1} + \frac{.33}{(1+.10)^2} + \frac{.33}{(1+.10)^3} + \frac{.33}{(1+.10)^4} + \frac{.33}{(1+.10)^5} + \frac{.31}{(1+.10)^6} + \frac{.28}{(1+.10)^7} \right] - 1.1 \right) = 0.47$$

$$FNPV_2 = \left(\left[\frac{.35}{(1+.08)^1} + \frac{.35}{(1+.08)^2} + \frac{.35}{(1+.08)^3} + \frac{.35}{(1+.08)^4} + \frac{.35}{(1+.08)^5} + \frac{.33}{(1+.08)^6} + \frac{.30}{(1+.08)^7} \right] - 1.0 \right) = 0.78$$

$$FNPV_3 = \left(\left[\frac{.37}{(1+.07)^1} + \frac{.37}{(1+.07)^2} + \frac{.37}{(1+.07)^3} + \frac{.37}{(1+.07)^4} + \frac{.37}{(1+.07)^5} + \frac{.35}{(1+.07)^6} + \frac{.31}{(1+.07)^7} \right] - .08 \right) = 1.14$$

$\therefore FNPV_{project\,I} = (0.47, 0.78, 1.14)$

Similarly $FNPV_{project\ II}$ and $FNPV_{project\ III}$ are (0.22, 0.65, 1.00) and (0.08, 0.38, 0.71) respectively.

The fuzzy payback period (in years) for the three projects are (3, 3, 4), (3, 4, 5) and (4, 5, 5) respectively.

Step 4

Sensitivity analysis is performed for three variables: expected demand, project life and discount rate. Expected demand is revealed through its effect on the annual net cash flows after tax. The annual net cash flows after tax are estimated to vary from between -15% to +15% of the original estimates. The fuzzy net present values calculated within this range are shown in table 7.8 and figures 7.9a-c. Similarly, the discount rate may also vary from between -15% to +15% of the original estimate. Resulted fuzzy net present values within this range are shown in table 7.9 and figures 7.10a-c. Finally, fuzzy net present values are calculated for project lives of 4, 5, and 6 years in addition to the original estimate of 7 years. These are shown in table 7.10 and figures 7.11a-c.

It can be seen from table 7.8 and figures 7.9a-c that under all changes in the annual net cash flows after tax, project I maintains a positive *FNPV*, while decreasing the annual net cash flows after tax by 15% leads to a negative *FNPV* for project II, and decreasing the annual net cash flows after tax by only 10% leads to a negative *FNPV* for project III. Table 7.9 and figures 7.10a-c indicate that the *FNPV* is not sensitive to changes in the discount rate for all three alternative projects. For the changes in the project life (see table 7.10 and figures 7.11a-c), the *FNPV* becomes negative if the life of project I is shortened to 4 years while it becomes negative if the

Table 7.8 Sensitivity of FNPV to changes in the annual net cash flows after tax (£000,000s)

	Project I	Project II	Project III
Original estimates	(0.47, 0.78, 1.14)	(0.22, 0.65, 1.00)	(0.08, 0.38, 0.71)
+ 5%	(0.55, 0.87, 1.24)	(0.32, 0.76, 1.11)	(0.18, 0.48, 0.83)
+ 10%	(0.63, 0.96, 1.34)	(0.41, 0.86, 1.22)	(0.28, 0.59, 0.94)
+ 15%	(0.71, 1.05, 1.43)	(0.50, 0.96, 1.33)	(0.38, 0.70, 1.06)
- 5%	(0.39, 0.69, 1.05)	(0.13, 0.55, 0.89)	(-0.02, 0.27, 0.60)
- 10%	(0.31, 0.60, 0.95)	(0.04, 0.45, 0.78)	(-0.12, 0.16, 0.48)
- 15%	(0.23, 0.51, 0.85)	(-0.05, 0.34, 0.67)	(-0.22, 0.05, 0.36)

Table 7.9 Sensitivity of *FNPV* to changes in the discount rate (£000,000s)

	Project I	Project II	Project III
Original estimates	(0.47, 0.78, 1.14)	(0.22, 0.65, 1.00)	(0.08, 0.38, 0.71)
+ 5%	(0.44, 0.76, 1.12)	(0.19, 0.63, 0.98)	(0.05, 0.34, 0.68)
+ 10%	(0.42, 0.73, 1.10)	(0.19, 0.60, .95)	(0.02, 0.32, 0.65)
+ 15%	(0.40, 0.71, 1.07)	(0.14, 0.57, 0.92)	(-0.01, 0.29, 0.63)
- 5%	(0.50, 0.80, 1.17)	(0.25, 0.68, 1.03)	(0.11, 0.40, 0.74)
- 10%	(0.52, 0.83, 1.19)	(0.28, 0.71, 1.06)	(0.15, 0.43, 0.77)
- 15%	(0.55, 0.85, 1.21)	(0.32, 0.74, 1.09)	(0.18, 0.47, 0.80)

Table 7.10 Sensitivity of FNPV to changes in the project life (£000,000s)

	Project I	Project II	Project III
7 years	(0.47, 0.78, 1.14)	(0.22, 0.65, 1.00)	(0.08, 0.38, 0.71)
6 years	(0.33, 0.61, 0.95)	(0.05, 0.44, 0.75)	(-0.11, 0.14, 0.45)
5 years	(0.15, 0.40, 0.72)	(-0.16, 0.20, 0.48)	(-0.34, -0.12, 0.16)
4 years	(-0.05, 0.16, 0.45)	(-0.40, -0.08, 0.19)	(-0.60, -0.41, 0.14)

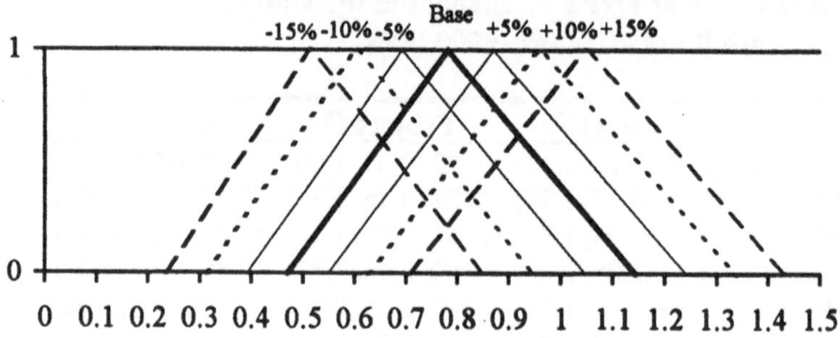

Figure 7.9a Sensitivity of *FNPV* to changes in the annual net cash flows after tax (project I)

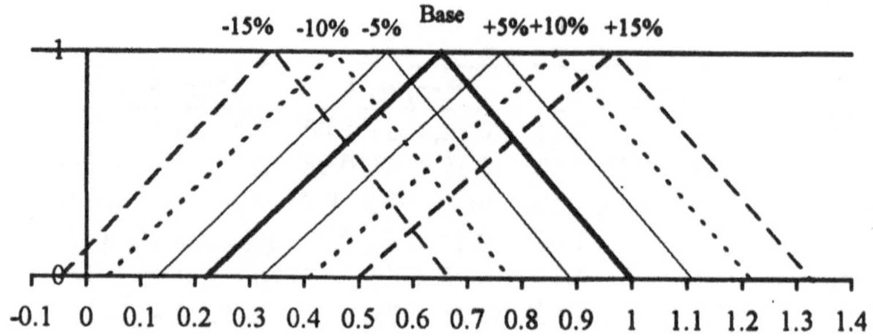

Figure 7.9b Sensitivity of *FNPV* to changes in the annual net cash flows after tax (project II)

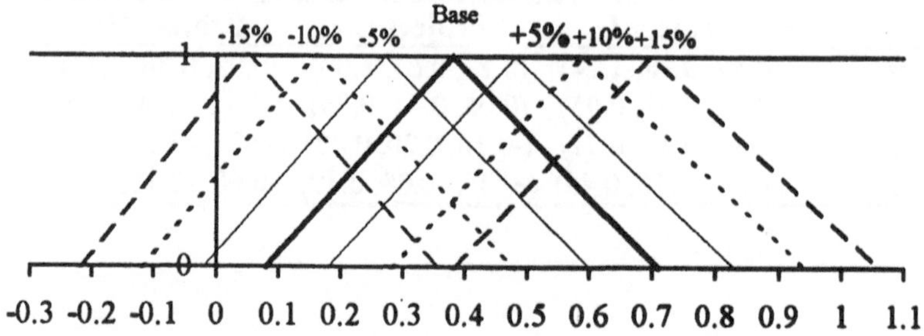

Figure 7.9c Sensitivity of *FNPV* to changes in the annual net cash flows after tax (project III)

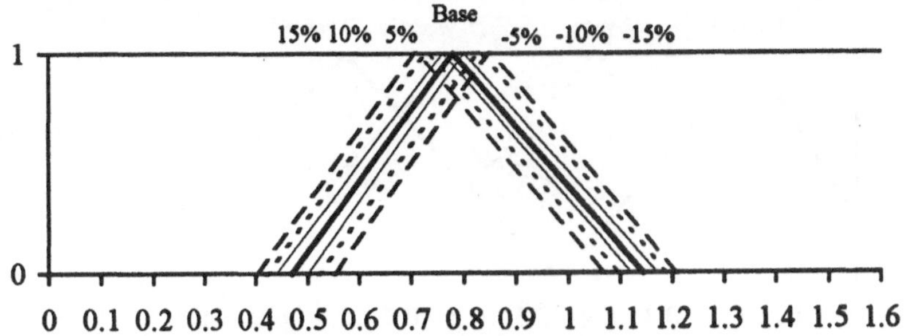

Figure 7.10a Sensitivity of *FNPV* to changes in the discount rate (project I)

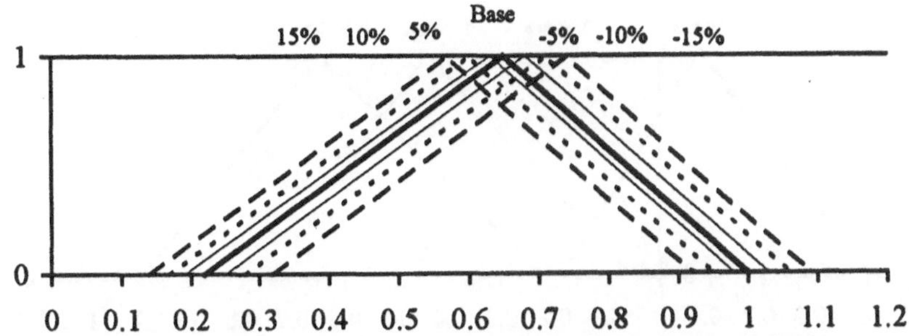

Figure 7.10b Sensitivity of *FNPV* to changes in the discount rate (project II)

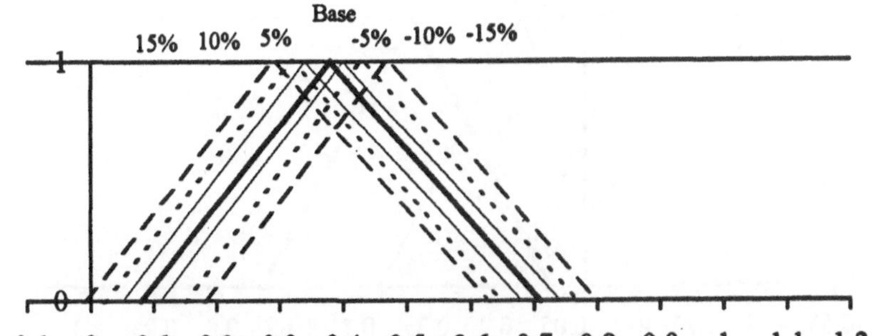

Figure 7.10c Sensitivity of *FNPV* to changes in the discount rate (project III)

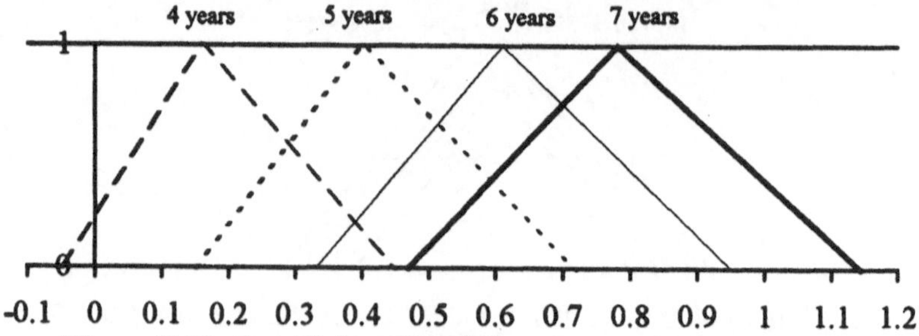

Figure 7.11a Sensitivity of *FNPV* to changes in the project's life (project I)

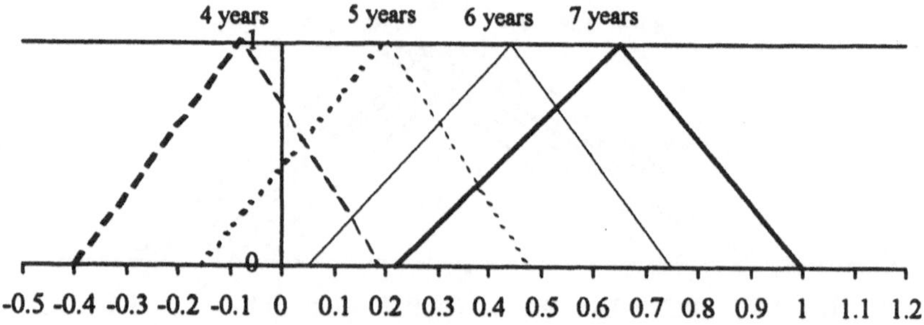

Figure 7.11b Sensitivity of *FNPV* to changes in the project's life (project II)

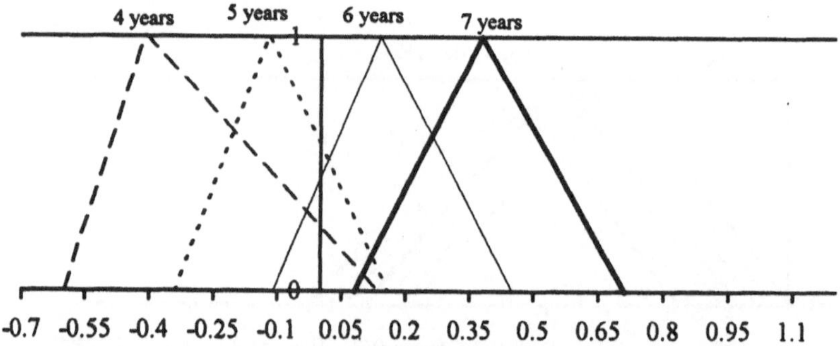

Figure 7.11c Sensitivity of *FNPV* to changes in the project's life (project III)

project life of project II is shortened by two years, and if the life of project III is shortened by only one year.

Step 5

The non-financial criteria are classified into three groups 'product quality', 'process flexibility' and 'customer requirements' as shown in figure 7.12.

Step 6

The decision makers employ the '*importance*' fuzzy linguistic scale of table 7.4.

Step 7

The importance of each non-financial criterion is determined qualitatively in table 7.11 and converted into its fuzzy number equivalent in table 7.12.

Step 8

The ratings assigned to each alternative with respect to the non-financial criteria are determined qualitatively as in table 7.13, and converted into their numerical equivalent[12] in table 7.14.

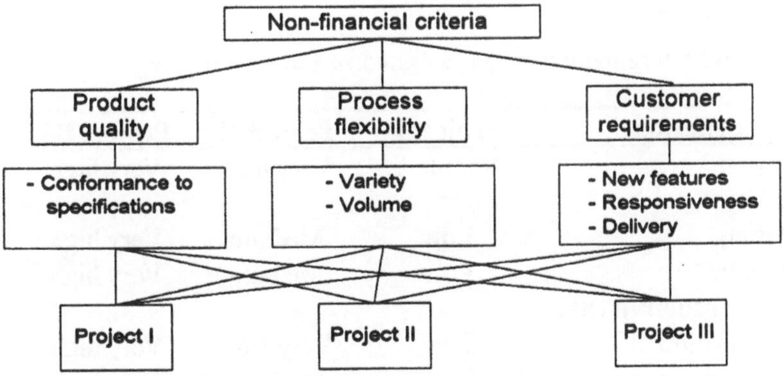

Figure 7.12 The non-financial criteria hierarchy

Table 7.11 The linguistic importance of the non-financial criteria

Criterion	Importance (Ci)
Quality	Very important
Flexibility	
Variety	Very important
Volume	Very important
Customer requirements	
New feature	Very important
Responsiveness	Important
Delivery	More-or-less important

Table 7.12 The scale importance of the non-financial criteria

Criterion	Importance (C_i)
Quality	(0.69, 1.00, 1.00)
Flexibility	
Variety	(0.69, 1.00, 1.00)
Volume	(0.69, 1.00, 1.00)
Customer requirements	
New feature	(0.69, 1.00, 1.00)
Responsiveness	(0.30, 0.45, 0.70)
Delivery	(0.16, 0.24, 0.37)

Table 7.13 The linguistic ratings assigned to the alternatives

Criterion	Project I	Project II	Project III
Quality	Very low	Very high	Very high
Flexibility			
Variety	Low	Medium	Very high
Volume	Low	Medium	Very high
Customer requirements			
New feature	Low	Very high	Very high
Responsiveness	Low	Medium	Very high
Delivery	Low	Medium	Very high

Table 7.14 The scale ratings assigned to the alternatives

Criterion	Project I	Project II	Project III
Quality	(0.00, 0.10, 0.19)	(0.70, 1.00, 1.00)	(0.70, 1.00, 1.00)
Flexibility			
Variety	(0.15, 0.25, 0.40)	(0.30, 0.50, 0.70)	(0.70, 1.00, 1.00)
Volume	(0.15, 0.25, 0.40)	(0.30, 0.50, 0.70)	(0.70, 1.00, 1.00)
Customer requirements			
New feature	(0.15, 0.25, 0.40)	(0.70, 1.00, 1.00)	(0.70, 1.00, 1.00)
Responsiveness	(0.15, 0.25, 0.40)	(0.30, 0.50, 0.70)	(0.70, 1.00, 1.00)
Delivery	(0.15, 0.25, 0.40)	(0.30, 0.50, 0.70)	(0.70, 1.00, 1.00)

Step 9

Using equation 7.31, the fuzzy non-financial measure for project I ($FNFM_{project\,I}$) can be calculated as follows:

$$W_{project\,I1} = 1/6\ [(.69)(.00) + (.69)(.15) + (.69)(.15) + (.69)(.15) + (.3)(.15) + (.16)(.15)] = 0.06$$

$$W_{project\,I2} = 1/6\ [(1)(.10) + (1)(.25) + (1)(.25) + (1)(.25) + (.45)(.25) + (.24)(.25)] = 0.17$$

$$W_{project\,I3} = 1/6\ [(1)(.19) + (1)(.40) + (1)\ (.40) + (1)(.40) + (.7)(.40) + (.37)(.40)] = 0.30$$

$\therefore\ FNFM_{project\,I} = (0.06, 0.17, 0.30)$

Similarly, $FNFM_{project\,II}$ and $FNFM_{project\,III}$ are (0.25, 0.56, 0.69) and (0.38, 0.78, 0.85) respectively.

Step 10

The risk factors are classified into two groups 'market risk' and 'technology risk' as shown in figure 7.13.

Step 11

Weighting the importance of risk factors can be undertaken using the fuzzy linguistic scale set out in table 7.4 while the fuzzy linguistic scale in table

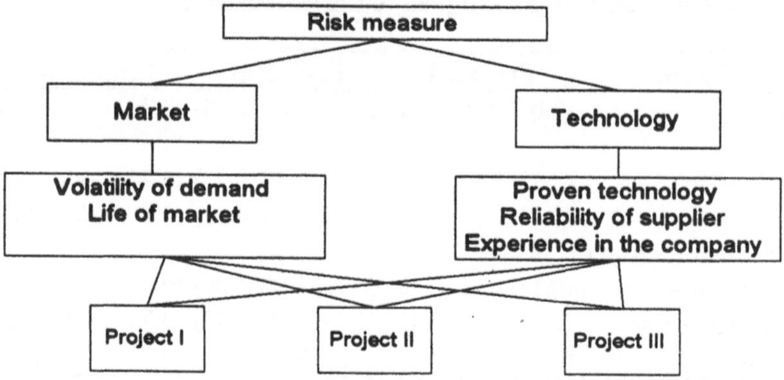

Figure 7.13 The risk measure hierarchy

Table 7.15 The importance weight of the risk factors

Factor	Weight
Volatility of demand	More-or-less important
Life of market	Important
Proven technology	Very important
Reliability of supplier	Very important
Experience in the company	More-or-less unimportant

Table 7.16 The equivalent fuzzy number weight of the risk factors

Factor	Weight
Volatility of demand	(0.16, 0.24, 0.37)
Life of market	(0.30, 0.45, 0.70)
Proven technology	(0.69, 1.00, 1.00)
Reliability of supplier	(0.69, 1.00, 1.00)
Experience in the company	(0.09, 0.11, 0.19)

7.5 can be used to evaluate each alternative with respect to each risk factor. Hence, the weight of the importance of each risk factor is determined qualitatively in table 7.15 and its equivalent fuzzy number is shown in table 7.16. The evaluation of each alternative with respect to each risk factor are determined qualitatively in table 7.17 and their equivalent fuzzy numbers are shown in table 7.18.

Step 12

Using equation 7.35, the fuzzy risk measure for project I ($FRM_{project\,I}$) can be calculated as follows:

$$FRM_{project\,I1} = 1/5\,[(.16)(.30) + (.3)(.16) + (.69)(.30) + (.69)(.09) + (.09)(0)] = 0.07$$

$$FRM_{project\,I2} = 1/5\,[(.24)(.45) + (.45)(.24) + (1)(.45) + (1)(.11) + (.11)(.06)] = 0.16$$

$$FRM_{project\,I3} = 1/5\,[(.37)(.70) + (.7)(.37) + (1)(.70) + (1)\,(.19) + (.19)(.08)] = 0.28$$

Table 7.17 The qualitative evaluation of the alternatives with respect to risk factors

Factors	Project I	Project II	Project III
Volatility of demand	Average	Very low	Non
Life of market	Long	Infinite	Infinite
Proven technology	Low	High	High
Reliability of supplier	High	Average	Very high
Experience in the company	Excellent	Good	Very poor

Table 7.18 The scale evaluation of the alternatives with respect to risk factors

Factors	Project I	Project II	Project III
Volatility of demand	(.30, .45, .70)	(.09, .11, .19)	(.00, .06, .08)
Life of market	(.16, .24, .37)	(.00, .06, .08)	(.00, .06, .08)
Proven technology	(.30, .45, .70)	(.09, .11, .19)	(.09, .11, .19)
Reliability of supplier	(.09, .11, .19)	(.16, .24, .37)	(.00, .06, .08)
Experience in the company	(.00, .06, .08)	(.16, .24, .37)	(.80, 1.0, 1.0)

$\therefore FRM_{project\,I} = (0.07, 0.16, 0.28)$

Similarly, $FRM_{project\ I2}$ and $FRM_{project\ I3}$ are (0.04, 0.09, 0.15) and (0.03, 0.06, 0.11) respectively.

Step 13

The summary of the three fuzzy measures computed for the three projects are presented in table 7.19. Also, the membership function of the three measures are depicted in figures 7-14a-c.

Using equations 7.37 - 7.41 and assuming $O = 0.5$, the ranked values (RV_{FNPV}), (RV_{FNFM}) and (RV_{FRM}) for *FNPV* , *FNFM* and *FRM* are calculated and shown in table 7.20. For example, the calculation of (RV_{FNPV}) is as follows:

$S = \{0.08, 0.22, 0.38, 0.47, 0.65, 0.71, 0.78, 1.00, 1.14\}$

$x_{max} = 1.14$

$x_{min} = 0.08$

$$RV_{FNPV}\text{(project I)} = (.78)(.5[(1.14-.08)/(1.14-.08+1.14-.78)] + (.5)[1-(1.14-.47)/(1.14-.08+.78-.47]) = 0.49$$

$$RV_{FNPV}\text{(project II)} = (.65)(.5[(1-.08)/(1.14-.08+1-.65)] + (.5)[1-(1.14-.22)/(1.14-.08+.65-.22]) = 0.34$$

$$RV_{FNPV}\text{(project III)} = (.38)(.5[(.71-.08)/(1.14-.08+.71-.38)] + (.5)[1-(1.14-.08)/(1.14-.08+.38-.08)]) = 0.13$$

Table 7.19 Summary of the three fuzzy measures

Fuzzy measures	Project I	Project II	Project III
FNPV	(0.47, 0.78, 1.14)	(0.22, 0.65, 1.00)	(0.08, 0.38, 0.71)
FNFM	(0.06, 0.17, 0.30)	(0.25, 0.56, 0.69)	(0.38, 0.78, 0.85)
FRM	(0.07, 0.16, 0.28)	(0.04, 0.09, 0.15)	(0.03, 0.06, 0.11)

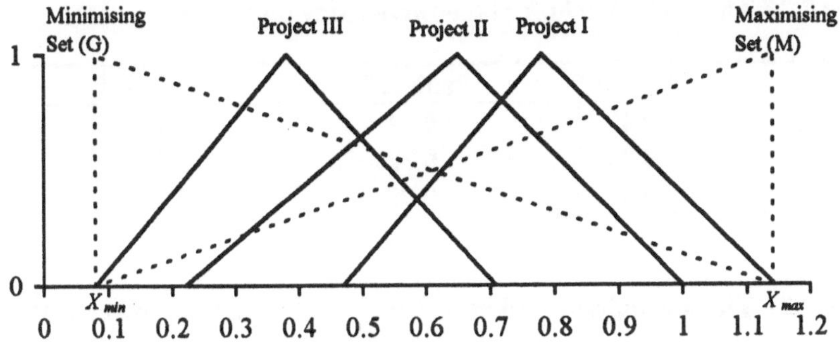

Figure 7.14a The *FNPV* for the three projects

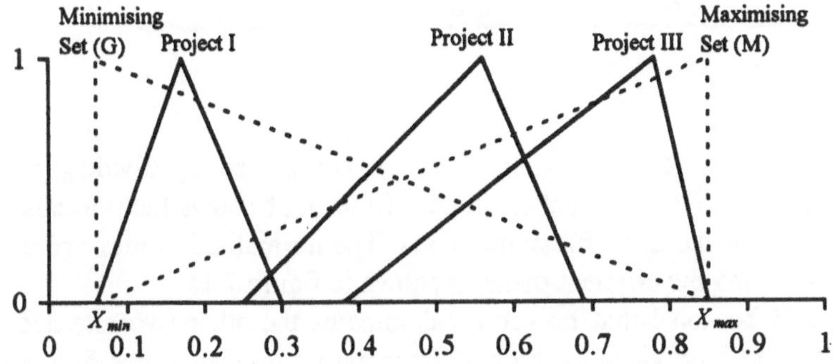

Figure 7.14b The *FNFM* for the three projects

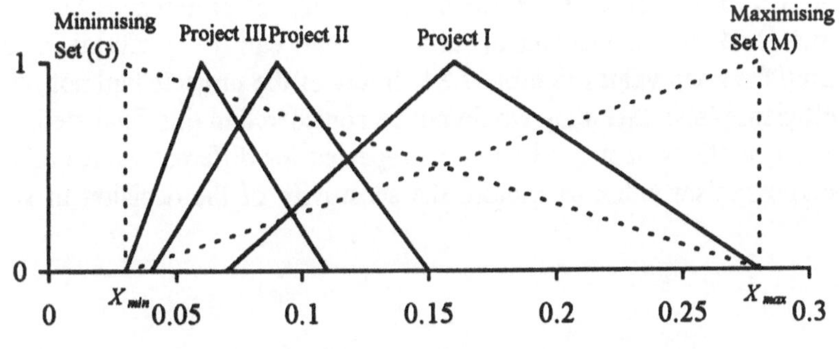

Figure 7.14c The *FRM* for the three projects

Table 7.20 The ranked values of the three fuzzy measures

	Project I	Project II	Project III
RV_{FNPV}	0.49	0.34	0.13
RV_{FNFM}	0.03	0.32	0.59
RV_{FRM}	0.08	0.03	0.01

Table 7.21 The normalised ranked values of the three fuzzy measures

	Project I	Project II	Project III
RV_{FNPV}	0.51	0.35	0.14
RV_{FNFM}	0.03	0.34	0.63
RV_{FRM}	0.67	0.25	0.08

Step 14

The ranked values of each measure can be normalised by dividing each ranked value by the total of ranked values. Table 7.21 shows the normalised ranked values for the three fuzzy measures. The normalised values are then used in developing the efficient plane as shown in figure 7.15.

It should be noted that no project dominates the other two alternative projects. Project I provides the highest financial return, but it has few non-financial benefits and is very risky. Project III provides many non-financial benefits and has the lowest risk but also the lowest NPV. Project II is in-between the other two projects with respect to the three measures. The selection of a project depends on the decision makers' preferences between NPV, risk and the non-financial criteria, and can be undertaken by considering the crisp values in table 7.21. If any of the projects had not been on the efficient plane, then they would not be considered in this final step.

The calculations in step 13 could be repeated for different values of the optimism/pessimism index to explore the sensitivity of the decision in step 14.

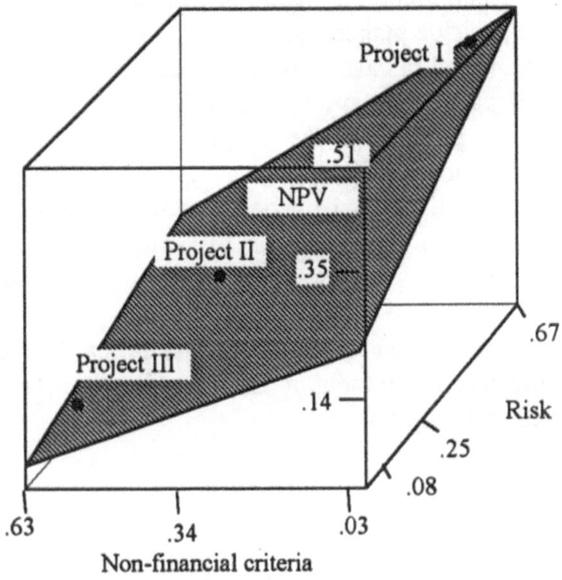

Figure 7.15 A graphical presentation of the example

Conclusions

In this chapter a model for AMT investment decision making has been developed. The model is based on fuzzy set theory, and unlike existing analytical models, the framework is based on the results of an empirical study. The model reduces the need for precise estimates that are difficult, if not impossible, to obtain in real-life situations and decision makers are enabled to express their estimates in approximated values and/or verbal values. Finally, the model provides decision makers with a formal trade-off in which they express their preferences between the financial return, non-financial benefits and multiple risk aspects of the AMT investment projects that are on the efficient plane.

The proposed model thus does not provide a single measure that combines all factors, rather a distinction is drawn between factors that can generally be quantified and factors that can only be qualitatively assessed. Accola's suggestion of a three-dimensional framework was followed in which a separate measure was developed for each dimension: financial return, non-financial benefits and risk, and the final decision is therefore a trade-off between the decision makers' preferences. However, if all benefits can be reasonably quantified in cash flow terms, then the three-dimensional framework can be collapsed into two-dimensions: financial return and risk.

The computational process in the proposed model was illustrated through a numerical example. Following a stepwise AMT investment decision algorithm. Although the model is based on fuzzy set theory, there is no need to be familiar with such calculus to perform the computations because the model's equations are designed in such a way that normal calculus is used.

Although the model presented in this chapter is primarily designed for AMT investment projects evaluation, it can, equally, be applied in the evaluation of other capital projects.

Notes

1 Derivation of the membership function of the fuzzy present value is deemed to be very important when decision makers have mutually exclusive projects that they need to rank in order to choose the most attractive. Most ranking methods (see chapter 6) depend on the membership function of every fuzzy number.

2 It should be noted that Wilhelm and Parsaei (1991) show these membership functions in four tables: primary values of variable X, primary values of variable Y, combined primary and hedge values of variable X, and combined primary and hedge values of variable Y. Presenting all the values in a single table revealed some relationships between the values and enabled the researchers to discover inconsistent membership functions between some values.

3 Choosing a suitable strategy depends on many factors which will not be discussed here. We simply consider if a proposed investment project is consistent with the strategy or not. (For details of the strategy process, see for example, Mintzberg et al., 1995 or Mintzberg, 1994).

4 This example is elicited from the analysis provided by Naik and Chakravarty (1992).

5 It should be noted that when applying routine fuzzy number 'arithmetic' in arriving at a financial measure, independent relationships between factors are assumed. For example, in the multiplication of two triangular fuzzy numbers, (a, b, c) and (d, e, f), each value in the first fuzzy number is multiplied by the corresponding value in the second fuzzy number, that is, minimum with minimum, maximum with maximum and so on (see chapter 6). If these two fuzzy numbers represent, for example, sales volume and unit price, then deriving revenue by their multiplication implies that when sales volume is at its minimum value, unit price is, also, at its minimum value. However, it might be expected that with lower unit price, sales volume would be enhanced. This expected inverse correlation is ignored in 'fuzzy arithmetic'.

6 This point was discussed in chapter 4.

7 When combining two or more financial return measures together, two problems should be addressed. First, the indication of each measure must be in the same direction, in other words, if for some measures the alternative with the highest (largest) value is the best, which for others it is the worst, summation will make no sense. In such circumstances all measures should be transformed to indicate the same direction. For example, the alternative with the smallest payback period is the most desirable, while the alternative with the highest NPV is the most desirable. In such a case, one of the two measures should be transformed to convey the same direction in the combined result. Second, All combined measures should be represented by same measurement unit. Because NPV gives values in pounds, IRR and ROI give their results in percentages, and the payback period is represented in years, all should be transformed into a score in the interval [0,1]. One possible way to overcome these two problems is to convert the preference of the payback period direction to be that the highest score is the best, and then to convert all original measures results into score in the interval [0,1] as follows:

For fuzzy payback measure:

$$FPBA_k^{\backslash} = (FPBA_k^{-1} / \sum_i^k FPBA_k^{-1}) / a$$

where:

$FPBA_k^{\backslash}$: the converted fuzzy payback period for the project A_k (k=1,2,...., N project),

$FPBA_k^{-1}$: the inverse of the payback period for the project i,

a : suitable constant so that all supports of $PBA_k^{\backslash}$ are in the closed interval [0,1].

For other measures (for example *FNPV*):

$$FNPVA_k^{\backslash} = (FNPVA_k / \sum_{k=1}^{N} NPVA_k) / a$$

where:

$FNPVA_k^{\backslash}$: the converted *FNPV* for the project A_k (k =1,2,...., N project),
$FNPVA_k$: the *FNPV* for the project A_k,
a : suitable non-fuzzy constant so that all supports of $NPVA_k\backslash$ are in the closed interval [0,1].

After converting all financial measures, they can be transformed into a single measure by weighting each measure and then combining them in the same way as for non-financial criteria measures.

8 It should be noted that these criteria are chosen for inclusion in the model because practitioners considered them to be very important when making investment decisions. One might ask why practitioners' views are privileged over those of academic researchers. The response would be simply that academic researchers' justification for the variables included in their models was found weak or non-existent. Practitioner's views are informed by their experiences in dealing with practical decision making problems. However, any criterion included in the model could be excluded or redefined to suit a specific context, and other criteria could be included.

9 The geometric mean method for assessing weight from a positive reciprocal matrix was extended to the fuzzy case by Buckley (1985). The only difference is that Buckley (1985) used trapezoidal fuzzy numbers while the triangular fuzzy numbers are used here.

10 The actual value is 1.43 which should be approximated to the highest value in the scale, i.e. 1.00, since any value more than the upper bound of the scale is redundant.

11 These criticisms have been discussed in chapters 1 and 2.

12 See table 7.4.

8 Summary and Conclusions

This final chapter draws together the key strands developed throughout the book. The purpose is to give the reader an overview of the AMT investment decision model and to highlight major issues identified. The review follows the same sequence as the book itself: a review and synthesis of previous decision models; the empirical studies (survey and field study); an examination of fuzzy set theory and the development of an AMT investment decision model. Finally, some suggestions are made that might guide future researchers in the field.

A Synthesis of Previous Models

Traditional investment models based on profit and/or cash flow analysis for the evaluation of AMT investments have been criticised for two reasons: the difficulty of quantifying the intangible benefits of AMT investments, and the problem of measuring and incorporating project risk into the investment model. Capital investment decision making theory has a long history and is continually developing paths in search of solutions to such problems. In this book, the literature reviewed was divided into two major strands: a normative (theoretical) strand and a descriptive (empirical) strand. In the normative literature, researchers have developed models which prescribe how decisions to invest in AMT *should* be made, while in the descriptive literature, researchers have investigated the practice of investment decision making in order to reach conclusions about how investment decisions are *actually* made. Both approaches can be criticised. The theoretical models developed in the normative or prescriptive literature often do not seem to find favour with practitioners, while the descriptive analyses often offer no guidance to practitioners as to how their practices might be improved. One may speculate that a third approach might be deemed to be more relevant: an approach rooted in practice but embodying normative ideas and concepts in order to improve practice. It is this third approach that has underpinned the central thrust of this book.

In chapter 2, existing normative AMT investment decision making models were critically reviewed. Such models usually concentrate on the problem of quantifying intangible benefits rather than dealing with risk issues. Possibly this is because measuring and incorporating risk into the decision model is a common problem in most investment decisions, while the quantification of intangible benefits is often related specifically to AMT investments. The literature was classified into three strands according to whether models concentrated on quantitative, qualitative or integrated methods. This classification facilitated the identification of factors included, excluded and any duplicated in the models reviewed.

Researchers in the quantitative tradition advocate that AMT investment decisions should be based on cash flow analysis. Conversely, researchers in the qualitative tradition suggest that AMT investment decisions should be regarded as 'strategic' and thus should be implemented regardless of the results of cash flow analysis. Both quantitative and qualitative approaches were reviewed and criticised on the grounds that both underestimate the importance of factors that can influence the decision.

Many researchers have attempted to reconcile the quantitative and qualitative approaches by building 'integrated' models that can incorporate both sets of factors. Relatively unsophisticated integrated models are based on a simple combination of multiple scores so as to generate an overall project score. However, as set out in the literature, such simple models are very sensitive to the subjectively assigned scores for qualitative variables, and so, for this and for other reasons, a number of researchers have suggested more sophisticated techniques, especially the analytic hierarchy process (AHP). The AHP has at least two advantages over simpler scoring methods. First, it incorporates an intuitively appealing hierarchical approach to the identification of the key factors in the decision. Second, it reduces subjectivity in the assignment of factor weightings by allowing decision maker(s) to compare only pairs of factors/alternatives in turn without attempting to set absolute weightings. A consistency checking procedure ensures that the weightings assigned are internally consistent. A number of researchers have adopted the AHP for modelling investment decision making, and in the view of the authors, Accola's (1994) AHP model represents the most cogently argued model as it extends the traditional return-risk framework to include a third dimension for intangible benefits.

Despite the relative sophistication of AMT investment decision models they can be criticised on two main grounds. First, the techniques embodied in the models require decision makers to provide *precise* values for the variables included in the decision model. Such precision is unlikely to reflect the vague, ambiguous and uncertain estimates made in practice which suggest the use of fuzzy set theory. Second, many models lack sufficient justification for the inclusion of some factors and the exclusion of others; a review of these models revealed an almost total lack of consistency. Overcoming this criticism was a key motivation for conducting the empirical studies reported in chapters 4 and 5.

The Empirical Studies

Although numerous models have been suggested in the normative literature, only limited empirical research has been undertaken although calls for further empirical investigation have been made by many authors (for example: Ho and Pike, 1991; Klammer et al., 1991, Slagmulder, 1995). The main aims of the empirical studies were to explore current practices used in evaluating AMT investment decisions and to identify the major factors affecting these decisions. The exploration provided both a better understanding of investment decision practices and a grounded basis for the factors that 'should' be included in AMT investment decision making.

Unlike previous studies, empirical data was collected using a triangulation of two types of methodology: a cross-sectional survey and a field study. The cross-sectional survey facilitated collection of quantifiable data from three groups of companies: companies that did not invest in AMT, companies that invested in less-integrated systems, and companies that invested in fully-integrated systems. Differences between the three groups of companies could thus be observed. The survey also facilitated access to practitioners by asking respondents to the questionnaire if they would be willing to participate in interviews. The triangulation of two research methodologies meant that validity and reliability of the research results were enhanced and that the limitations of each research methodology were minimised. Triangulation was adopted for two main reasons. First, a sufficiently large sample of companies would be surveyed in order to identify the key factors affecting AMT investment decisions and to determine whether there were differences in the importance of such factors across companies employing different levels of AMT. Second, a field study would allow explanation of some of the survey results, and more importantly, would permit a description of how AMT investment decisions are actually made.

The cross-sectional survey reported in chapter 4 had a number of novel aspects compared with previous surveys:

1. Most of the qualitative factors reported in the capital investment literature were investigated, whereas in previous surveys, only a few such factors were considered (Pike, 1989) or the factors were treated as a single group (Chen, 1995), or they were not considered at all.
2. Unlike in most previous surveys, in which respondents were asked to report their use of capital budgeting and risk analysis techniques, respondents were asked to rate the *importance* of each technique or criterion on a five-point scale. Previous surveys had indicated that most companies use more than one technique when evaluating a capital investment decision although this does not necessarily mean that all the techniques are given equal importance in decision making.
3. This survey contrasted results between AMT and non-AMT companies, and distinguished between degrees of integration in AMT.
4. While most previous surveys only reported simple descriptive statistics, here statistical comparisons were based on Kruskal-Wallis one-way ANOVA, and so the relative importance of various techniques and criteria across the three groups of companies could be ascertained.

The purpose of the field study was to obtain descriptions of specific AMT investment decisions as they were taken. This contrasts with previous studies that investigated the impact of investment decision making on predetermined issues. For example, Slagmulder and Bruggeman (1992) investigated the relation between different investment justification methods and subsequent project financial performance.

Unstructured interviews were adopted as the primary method in the field study. These involved as little prompting of respondents as possible during the description of a specific investment decision. Triangulation was achieved within each interview by following up with a checklist of questions.

The survey confirmed earlier conclusion of the literature that investment projects can be affected by three main groups of factors: financial return, non-financial factors, and risk. In the three groups of companies, while all financial return measures were important, the simpler measures of ROI and payback were marginally more important than the more sophisticated measures of NPV and IRR. Kruskal-Wallis one-way ANOVA suggested that there were no significant differences between the importance of each financial return measure across the three groups of companies. Thus, the sophistication of the investment project being evaluated does not seem to make a significant difference to the importance of the financial return measure(s) used.

AMT investments can be penalised by using stringent financial criteria (high discount rates and short payback periods), and by failing to capture many of the intangible benefits that should flow from them. This conclusion is based on three of the survey results. First, the survey indicated that 3 years or less was set as a payback target by most of the sample companies. Second, the majority of companies used a high discount rate of 16% or more. Finally, many intangible benefits were not considered in the evaluation process for AMT projects because they are not easily quantifiable. This was confirmed in two ways: by the agreement of the majority of the respondents to the statement that 'some of the potential benefits of AMT are not taken into account because they are difficult to express in financial terms', and only 7 potential benefits of the 21 listed in the survey were considered in financial terms by the majority of respondents, and that a significant number of respondents did not consider some of the benefits at all.

Interviewees confirmed the importance of all the financial return measures in investment decision making. More than one financial return measure was used usually in order to provide the decision maker(s) with more information and/ or to cope with the differing preferences of the multiple participants involved in the decision making process.

Non-financial criteria play an important role in the decision regarding both AMT and other investments. This importance was confirmed in two ways. First, in the survey, 11 non-financial criteria, usually cited in the literature as important factors in the investment decision, were listed and respondents were asked to rate the importance of each criterion using a five-point scale. Second, in the field study, interviewees were asked to indicate the important factors which affect investment decisions *without* having been prompting by specific criteria.

The survey results indicate that at least eight non-financial criteria were important in evaluating investment projects. They are: 'requirements of customers', 'consistency with corporate strategy', 'keeping-up with competition', 'quality and reliability of outputs', 'obtaining greater manufacturing flexibility', 'the ability to expand in the future', 'reduced lead-times', and 'reduced inventory levels'. Further, comparison of the importance of these non-financial criteria across the three groups of companies indicated that these factors were seen to be more important by the fully integrated companies than by the other two groups of companies (non-

AMT and less integrated companies). The higher importance of non-financial criteria in the fully integrated companies was statistically significant for four of the non-financial criteria: 'quality and reliability of outputs', 'obtaining greater manufacturing flexibility', 'reduced lead time', and 'reduced inventory levels'. These four criteria were seen to be increasingly important across the spectrum from non-AMT to less integrated to fully integrated AMT companies. It is concluded that these factors were often the driving forces behind AMT investment decisions while the other four non-financial factors: 'requirements of customers', 'consistency with corporate strategy', 'keeping with competition', and 'the ability to expand in the future' were important factors for investment decisions in general, no matter whether AMT or not.

In the field study, the importance of non-financial criteria in the investment decision was confirmed. However, there is little evidence that they were considered in a formal way. Scoring methods were only used in three cases and for limited purposes. Most non-financial factors were discussed in the final stage of the decision process. However, strategic implications were treated differently. In almost all cases, investment projects were reviewed for consistency with corporate strategy *before* conducting a detailed analysis.

Some factors have both financial and non-financial implications. For example, improvements in product quality usually result in reduced scrap, rework, and inspection costs as well as potentially leading to increased sales. However, they could also lead to improvements in company image and in the competitive position of the company. These benefits are hard to quantify in financial terms. Reduced lead-times can be partially quantified in financial terms because such reductions can lead to increased sales volume. However, other benefits, such as faster response to customer requirements giving the company an advantage over its competitors, are difficult to quantify financially. Reductions in inventory levels are most appropriately considered as part of the financial analysis because savings resulting from reductions in the costs of holding, moving, scheduling and inspecting inventories can be quantified.

The survey results indicated that the evaluation of the riskiness of AMT investments is a difficult task. Sensitivity analysis is the most important technique for dealing with investment risk in practice. This was confirmed in the field study by being routinely carried out and receiving considerable attention during the decision making process. The more sophisticated risk analysis techniques, such as probability analysis, simulation, and the capital asset pricing model were found to be not widely used.

Most of the respondents to the survey indicated that they use a short payback period (3 years or less) and a high discount rate (16% or more) in evaluating AMT investments. However, these targets are not applied specifically to AMT investments, similar criteria are employed for non-AMT investments. Although not statistically significant, AMT companies, tended to employ more stringent financial criteria than non-AMT companies, and this could militate against AMT investment, given the long-lived nature of such investments and the time-scale needed for the benefits to fully materialise. However, simple conclusions should not be drawn as respondents indicated that AMT investments were not particularly more difficult to approve than other categories of investment.

There was no evidence in the survey that financial criteria were set by formal analysis which might relate such criteria to the nature of the investment or to the perceived riskiness of the sector. This was confirmed in the field study in which interviewees claimed, typically, to use industry averages for financial criteria and hurdle rates.

Although it was difficult to obtain a clear idea about how practitioners deal with investment risk from the survey responses, this issue was further investigated in the field study. During the discussions with the interviewees it was clear that they were more interested in *minimising* the risk of investment projects than *measuring* it. Risk was usually analysed in terms of its sources, and efforts made to eliminate or minimise it. Two sources of risk were usually identified in AMT investments: market risk and technology risk. Examples were given by interviewees of how these risks could be reduced or eliminated, such as by sharing them with other parties, such as suppliers or customers. The idea of thinking about risk in terms of its sources supports the dimension-oriented (rather than moment-oriented) approach.

The empirical study concluded by presenting a general model for AMT investment decision making. The model is grounded in existing theory (Accola, 1994) and in current practice as revealed in previous surveys/field studies and in the empirical work reported here. The model therefore provides a sound basis for further theoretical developments utilising the mathematics of fuzzy set theory.

An Examination of Fuzzy Set Theory

Fuzzy set theory provides a means of dealing mathematically with fuzzy variables. Fuzzy variables can be assigned values in approximate forms such as 'about £60' and 'around 3 years' and/ or qualitative forms such as 'high' and 'low'. Fuzzy set theory is appropriate for modelling AMT investment decisions for two main reasons. First, all factors affecting the decision can be included whether they can be quantified or not; 'intangible' variables need not be excluded from analysis. Second, it reduces the need for precise inputs to the decision model by permitting imprecise estimates of the values of the variables.

'Fuzzy' is a term used to describe an imprecise quantity. Three special forms of fuzzy numbers have been suggested in the fuzzy set literature: L-R fuzzy numbers, triangular fuzzy numbers, and trapezoidal fuzzy numbers. By using fuzzy numbers and their algebraic operations, financial return measures (such as NPV and IRR) can be extended to the case where cash flows are expressed as imprecise values.

In addition to its use in defining approximate quantitative variables, fuzzy set theory has been extended so that linguistic variables can also be represented in fuzzy form. The value of a linguistic variable is represented by a linguistic form (word or sentence) rather than a number. Each such linguistic value is represented by a fuzzy subset in which the membership function represents the compatibility of each element, in the universe of discourse, with the label of the fuzzy subset. Thus, linguistic values can be dealt with mathematically by using fuzzy subsets as surrogates. By using the linguistic approach, the non-financial criteria of an investment decision can be expressed in linguistic values. The mathematics of fuzzy set theory can then be used to arrive at a measure for the non-financial criteria.

When the outcomes of multiple alternative choices are presented in fuzzy form, there may be difficulty in ranking them. This difficulty arises because each fuzzy number is represented by imprecise (multiple) quantities, and fuzzy numbers are therefore likely to 'overlap'. Ranking methods have been suggested in the literature, but different methods can lead to different rankings for the same fuzzy numbers. This is because the methods suggested utilise only partial information contained in each fuzzy number. A fuzzy number can be divided into three parts: members that have full memberships, members that have partial memberships located in the right-hand side, and members that have partial memberships located in the left-hand side. Existing ranking methods either reflect the membership functions of the left-hand side or of both sides. In order to reflect all three parts of a fuzzy number in the ranking process, a new method for ranking fuzzy numbers is proposed.

In the proposed ranking method, a ranking value is computed for each fuzzy number under consideration. This ranking value is a member of the fuzzy number, which has full membership, weighted by the average score of both sides of the membership function of the fuzzy number. If two fuzzy numbers have the same ranking values, the divergence (the difference between the lowest and highest values in the fuzzy number) is used to differentiate them, and the fuzzy number having the lowest divergence is ranked higher. Two fuzzy numbers are equal if and only if both have the same ranking value and divergence. The method also allows for degrees of optimism/pessimism in the calculation of the ranking value.

The Development of an AMT Investment Decision Model

The AMT investment decision model is based on two elements: a framework derived from the empirical study so that factors included in the model are justified by their practical importance, and the mathematics of fuzzy set theory. The model is designed to overcome the lack of a consistent method of accommodating non-financial factors, and should lead to better decisions as all factors affecting the decision are considered regardless of whether they are financial or non-financial. It also provides a framework for the comparison of competing alternatives so that the decision maker(s) can choose the most appropriate alternative.

Before developing a new model for AMT investment decision making based on fuzzy set theory, previous studies based on 'fuzzy' modelling of investment decisions were reviewed. Two approaches were identified. The first approach extends traditional discounted cash flow analysis into fuzzy cash flow analysis (Ward, 1985 and 1989; and Chiu and Park, 1994). The second approach is based on the concept of linguistic variables (Wilhelm and Parsaei, 1988 and 1991). However, previous studies have not combined both financial and linguistic variables into an integrated analysis.

Ward (1985 and 1989) extended the traditional calculation of NPV by allowing cash flows to take fuzzy form. Deriving the membership function of NPV according to Ward's suggestions requires tedious computational effort as it results in complex non-linear representations. However, Chiu and

Park's (1994) suggestion overcomes this problem by using an approximate form of the 'complex' fuzzy present value formula used by Ward. Based on simulation results, they concluded that the difference between the membership function of the present value formula in its exact and approximate forms is not significant as long as the width of the discount rate is within a range of ±4%.

The second approach, based on linguistic variables, was suggested by Wilhelm and Parsaei (1988 and 1991). They define two fuzzy linguistic variables 'importance' and 'capability'. The first variable is used to specify the relative importance of each goal within a set of common goals. The second variable is used to specify the capability of each technology in meeting each of the goals of the company. Each variable takes three primary linguistic values defined, subjectively, in the range [0,1]. Then, by using the fuzzy operations of 'negation', 'and' and 'or' and linguistic hedges, new values can be generated. Using these two fuzzy linguistic variables, Wilhelm and Parsaei (1988 and 1991) developed a heuristic algorithm to characterise the capability of available advanced technologies and so select the technology that best meets the company's goals according to their importance.

Although seminal in its application of fuzzy set theory to AMT investment decisions, Wilhelm and Parsaei's (1989 and 1991) work was criticised on three counts. First, the computations involved in constructing relationships between linguistic variables seem unjustifiably complex. Second, perhaps because of this complexity, some relationships between linguistic values were seen to be counter-intuitive and even inconsistent when gathered in a single table (table 7.1). Finally, financial factors were ignored in Wilhelm and Parsaei's model or, at most, treated as though they were intangibles.

It is evident that the AMT investment decision is affected by both financial and non-financial factors and therefore both fuzzy numbers and fuzzy linguistic variables should be used. A 'comprehensive' approach for modelling AMT investment decision based on both concepts (fuzzy numbers and linguistic variables) is proposed here.

The AMT investment decision can be divided into three sequential stages: examining the consistency of an AMT investment project with overall strategy, measuring the expected performance of alternatives, and ranking competing AMT technologies.

The first stage ensures that all investment projects are consistent with the corporate strategy so that only those projects which are consistent with overall strategy are subjected to detailed evaluation.

The second stage measures the expected performance of AMT investments based on three dimensions: financial return, intangible benefits, and risk. Financial return reflects all the factors that can be translated into cash flows, each factor being represented by a triangular fuzzy number. One or more financial measure: NPV, IRR, payback and/or ROI, can then be calculated. Traditional NPV was extended to the case of fuzzy cash flows by Ward (1985 and 1989) and Chiu and Park (1994), and suggested extensions for the other measures were presented in chapter 7. Decision makers can choose their preferred measure(s) to represent the financial return dimension in the model.

The non-financial criteria measure reflects all the factors that cannot be translated into cash flows. In order to provide a measure for this group of factors, the fuzzy linguistic approach is suggested. Previous studies which used the concept of linguistic variables in decision making usually suggested the use of generalised linguistic values. Here, the linguistic approach is used to generate a linguistic scale which should be *generated by the decision makers themselves*. This is important because the meaning (values) of any word (linguistic variable) depends mainly on the decision maker(s) involved in the decision. It is suggested that the linguistic scale is generated by using pair-wise comparisons between the linguistic values used in the evaluation process. By generating a linguistic scale based on pair-wise comparisons, the relative difference between each pair of linguistic values can be reflected.

The third dimension of the proposed model for AMT investment decision is project risk. The empirical results suggest that sensitivity analysis can be a first step in analysing project risk in order to identify the most sensitive variables and sources of risk. Possibilities to reduce or eliminate this risk should then be addressed. The next step, incorporated in the model, is to measure the residual risk quantitatively using fuzzy numbers or qualitatively using a linguistic scale. Two types of risk were discussed in the field study: market risk and technology risk. Both types can be represented as sub-dimensions of the risk dimension in the model. Then, based on the importance of each risk factor, they can be combined to generate an overall risk measure for each investment project.

After developing the three fuzzy measures of financial return, non-financial criteria, and risk, for each investment project under consideration, each fuzzy measure can be converted to an equivalent crisp measure so that the relative differences between competing investment projects can be compared. The conversion can be achieved by using the ranking method developed earlier. The crisp measures are then used in constructing a graphical representation and, according to the preference of the decision maker(s), the 'best' investment project is selected for implementation. The graphical representation of the problem follows Accola (1994).

An algorithm summarising the proposed model for AMT investment decision making was presented, and illustrated in the solution of a hypothetical AMT investment decision problem.

It can be seen therefore, that the proposed model does not provide a single measure that combines all factors, rather a distinction is drawn between factors that can be quantified and factors that can only be subjectively assessed. The final decision is thus subject to decision makers' preferences. If all the benefits can be reasonably quantified in cash flow terms, then the three-dimensional framework can be collapsed into two-dimensions: financial return and risk. Although the model is based on the fuzzy set theory calculus, there is no need to be familiar with it to perform the computational requirements of the model because the model's equations are designed in such a way that normal arithmetical operations are used.

All of the factors included in the model are based on the results of the empirical work which is the justification for their inclusion. One may ask why practitioners' views are privileged over those of academic researchers. The response is that practitioners' views are informed by their experience in dealing with practical decision making problems while academic researchers' justifications for the variables included in their models was found to be weak or non-existent. Practitioners are also more likely to be convinced by a model based on practical experience. However, any factor included in the model could be excluded or redefined to suit a specific context.

Finally, although the proposed model is primarily designed for AMT investment project evaluation, it can equally be applied to evaluate any capital project.

Suggestions for Further Research

The AMT investment decision model proposed in this book is founded on the assumptions of microeconomics in that those who would use the model are presumed to be rational economic actors intent on maximising long run wealth subject to some risk constraint (Davis and Pointon, 1994 and Ryan et al., 1992). The model can appear complex which might be a deterrent to its use in practice. There is considerable evidence that complex management accounting techniques are often little used. Indeed, there has been extensive discussion of 'the gap' between management accounting theory and practice during the past twenty years (see, for example, Scapens et al., 1996; Ryan et al., 1992; and Johnson and Kaplan, 1991)

Further research might logically address the issues which arise in implementing the model in practice. Two separate but linked strands of research suggest themselves. The first strand would address the computational aspects of the model. Although there is some complexity in the mathematics which underpins the model, it would be possible to develop software which automates the calculation aspects of the model. This research might be seen as converting the model into a 'black box' which allows the input of data, assumptions and value judgements, and produces the consequences of the data as the evaluation of the alternative projects. The development of such software might begin to address one of the implementation issues by reducing the perceived complexity of the model. One can see this process at work in a number of other areas such as linear programming, statistical analysis and even, relevant to this project, discounted cash flow analysis. In these areas 'user-friendly' software such as SPSS, Lotus 1-2-3 and Excel have made relatively complex techniques accessible to practitioners.

The reduction in the perceived complexity in these examples is not, however, achieved without cost. There is the danger that practitioners may use sophisticated techniques without understanding them, and may, therefore choose inappropriate methods of analysis. It is also possible that some practitioners reject appropriate methods either because they do not understand the technique or the results, or because they do not feel sufficiently 'involved' in the process.

The second stand of research would be to address such implementation issues. In the authors' view the model lends itself to an implementation path that could overcome resistance by developing understanding and confidence in the model.

First, the model could be introduced as providing an over-arching methodology for evaluating the return, risk and intangible characteristics of investment proposals. As each separate project dimension is evaluated separately, this allows the user to grasp the principles which underlie each element of analysis. Many practitioners will be familiar with financial analyses and this provides a firm starting point in AMT project evaluation.

The second step would be to employ the discipline of the model to force decisions on two crucial issues: how the user plans to deal with risk, and how to identify those issues which have been captured in the financial analysis and those which have not. One of the advantages of the suggested model is its ability to accommodate simplified assumptions, and here, for example, a user could simply specify a risk adjusted discount rate as the preferred approach in handling project risk. This would reduce the dimensions in the model which would reduce its sophistication but improve its understandability and ease of interpretation.

The third step would be to elicit value judgements concerning the relative importance of the intangible benefits and the relative performance of the alternative projects. The model has a number of advantages here which might make it attractive to practitioners:

1. A checklist of intangible benefits which are typically taken into account in the evaluation of AMT projects.
2. The analytic hierarchy structure which is intuitively appealing and provides a logical structure for evaluation.
3. The use of a linguistic scale in evaluating intangible benefits.

User understanding and confidence could be developed by building a simplified hierarchy in the first instance, perhaps including only those intangible variables which are considered to be especially important. This would allow the sensitivity of the addition of further intangible benefits to be readily evaluated.

In conclusion, it is the researchers' opinion that a logical implementation path exists which could reduce perceived complexity, develop understanding and, through the involvement of users, especially in the evaluation of intangibles, develop user commitment. It is well known that the successful use of techniques depends upon careful attention to the technical/human interface, and the model could provide a basis for further research in this area.

Bibliography

Abdel-Kader, M. (1997), *Evaluating Investment Decisions in Advanced Manufacturing Systems- A Fuzzy Set Theory approach*, Unpublished PhD Thesis, University of the West of England, Bristol.

Abdel-Kader, M., and Dugdale, D. (1996), *Investment Decisions in Advanced Manufacturing Technology*, In: Applied Accounting Research Series, Working Paper No. 5, University of the West of England, Bristol.

Abdel-Kader, M., and Dugdale, D. (1998), Investment in Advanced Manufacturing Technology: A Study of Practice in Large UK Companies, *Management Accounting Research*, Vol. 9, No. 3, pp.261-284.

ACARD (1983), *New Opportunities in Manufacturing: The Management of Technology*, Her Majesty's Stationery Office, London, October.

Accola, W. (1994), Assessing Risk and Uncertainty in New Technology Investments, *Accounting Horizons*, Vol. 8, No. 3, September, pp. 19-35.

Accola, W., Agrawal, S., and Holsapple, C. (1995), Adaptive Decision Support Systems for Evaluating Risk and Uncertainty in Capital Investment Decisions: Opportunities for Future Research, *Managerial Finance*, Vol. 21, No. 3, pp. 1-16.

Airey, J., and Young, C. (1983), Economic Justification-Counting the Strategic Benefits, *Proceedings of 2nd International Conference on Flexible Manufacturing Systems*, Rathmill, K. (editor), London, October.

Anderson, D., Sweeney, D., and William, T. (1994), *An Introduction to Management Science: Quantitative Approaches to Decision Making*, West Publishing Company, New York.

Angelis, D., and Lee, C. (1996), Strategic Investment Analysis Using Activity Based Costing Concepts and Analytical Hierarchy Process Techniques, *International Journal of Production Research*, Vol. 34, No. 5, pp. 1331-1345.

Apostolou, B., and Hassell, J. (1993), An Overview of the Analytic Hierarchy Process and its Use in Accounting Research, *Journal of Accounting Literature*, Vol. 12, pp. 1-28.

Asfahl, C. (1992), *Robots and Manufacturing Automation*, John Wiley & Sons, Inc., New York.

Ashenbrenner, K. (1984), Moment-Versus Dimension-Oriented Theories of Risky Choice: A (Fairly) General Test Involving Single-Peaked Preferences, *Journal of Experimental Psychology: Learning, Memory, and Cognition*, Vol. 10, No. 3, July, pp. 513-535.

Attaran, M. (1992), Flexible Manufacturing Systems: Implementing an Automated Factory, *Information Systems Management*, Vol. 9, No. 2, Spring, pp. 44-47.

Azzone, G., Bertele, U., and Masella, C. (1992), Selecting Techniques for the Financial Justification of Advanced Manufacturing Technologies: Contingent Approach, In: *Economic and financial Justification of Advanced Manufacturing Technologies*, Parsaei, H., Sullivan, W., and Hanley, T. (eds.), Elsevier Science Publishers, pp. 19-35.

Baldwin, J., and Pilsworth, B. (1982), Dynamic Programming for Fuzzy Systems with Fuzzy Environment, *Journal of Mathematical Analysis and Applications*, Vol. 85, No. 1, January, pp. 1-23.

Bell, D., Raiffa, H., and Tversky, A. (1988), Descriptive, Normative, and Prescriptive Interactions in Decision Making, In: *Decision Making: Descriptive, Normative, and Prescriptive Interactions*, Bell, D., Raiffa, H., and Tversky, A. (eds.), Cambridge University Press, Cambridge, pp. 9-30.

Bellman, R., and Zadeh, L. (1970), Decision-Making in a Fuzzy Environment, *Management Science*, Vol. 17, No. 4, pp. 141-164.

Bennett, R., and Hendricks, J. (1987), Justifying the Acquisition of Automated Equipment, *Management Accounting*, Vol. 67, July, pp. 39-46.

Boaden, R., and Dale, B. (1990), Justification of Computer-Integrated Manufacturing: Some Insights into the Practice, *IEEE Transactions on Engineering Management*, Vol. 37, No. 4, November, pp. 291-296.

Booth, L. (1996), Making Capital Budgeting Decisions in Multinational Corporations, *Managerial Finance*, Vol. 22, No. 1, pp. 3-17.

Boucher, T., and MacStravic, E. (1991), Multiattribute Evaluation Within a Present Value Framework and its Relation to the Analytic Hierarchy Process, *The Engineering Economist*, Vol. 37, No. 1, Fall, pp. 1-32.

Buckley, J. (1985), Fuzzy Hierarchical Analysis, *Fuzzy Sets and Systems*, Vol. 17, pp. 233-247.

Butler, R., Davies, L., Pike, R., and Sharp, J. (1991), Strategic Investment Decision-Making: Complexities, Politics and Processes, *Journal of Management Studies*, Vol. 28, No. 4, July, pp. 395-415.

Butler, R., Davies, L., Pike, R., and Sharp, J. (1993), *Strategic Investment Decisions: Theory, Practice and Process*, Routledge, London.

Canada, J., and Sullivan, W. (1990), Persistent Pitfalls and Applicable Approaches for Justification of Advanced Manufacturing Systems, *Engineering Costs and Production Economics*, Vol. 18, pp. 247-253.

Carr, C., and Tomkins, C. (1996), Strategic Investment Decisions: The Importance of SCM. A Comparative Analysis of 51 Case Studies in U.K., U.S. and German Companies, *Management Accounting Research*, Vol. 7, No. 2, June, pp. 199-217.

Carr, C., Tomkins, C., and Bayliss, B. (1994), *Strategic Investment Decisions: A Comparison of UK and German Practices in the Motor Components Industry*, Avebury, Aldershot.

Chan, Y., and Lynn, B. (1993), Hierarchical Analysis as a Means of Evaluating Tangibles and Intangibles of Capital Investments, *The Mid-Atlantic Journal of Business*, Vol. 29, No. 1, March, pp. 59-74.

Chan, Y., and Yuan, Y. (1990), Dealing with Fuzziness in Cost-Volume-Profit Analysis, *Accounting and Business Research*, Vol. 20, No. 78, pp. 83-95.

Chandra, J., and Schall, S. (1988), Economic Justification of Flexible manufacturing Systems Using the Leontief Input-Output Model, *The Engineering Economist*, Vol. 34, No. 1, Fall, pp. 27-50.

Charles Stark Draper Laboratory (1984), *Flexible Manufacturing System Handbook*, Noyes Publications, New Jersey.

Chen, I., and Small, M. (1996), Planning for Advanced Manufacturing Technology: A Research Framework, *International Journal of Operations and Production Management*, Vol. 16, No. 5, pp. 4-24.

Chen, S. (1995), An Empirical Examination of Capital Budgeting Techniques: Impact of Investment Types and Firm Characteristics, *The Engineering Economist*, Vol. 40, No. 2, Winter, pp. 145-170.

Chen, S., and Hwang, C. (1989), Fuzzy Scoring of Fuzzy Number- A Direct Comparison Index, Unpublished Paper, As Cited by Chen, Shu-Chen, and Hwang, C., in Collaboration with Hwang, F. (1992), *Fuzzy Multiple Attribute Decision Making: Methods and Applications*, Springer-Verlag, New York.

Chen, Shan-Huo (1985), Ranking Fuzzy Numbers with Maximizing Set and Minimizing Set, *Fuzzy Sets and Systems*, Vol. 17, No. 2, pp. 113-129.

Chen, Shu-Chen, and Hwang, C., in Collaboration with Hwang, F. (1992), *Fuzzy Multiple Attribute Decision Making: Methods and Applications*, Springer-Verlag, New York.

Cheung, J. (1993), Managerial Flexibility in Capital Investment Decisions Literature, *Journal of Accounting Literature*, Vol. 12, pp. 29-66.

Chiu, C., and Park, C. (1994), Fuzzy Cash Flow Analysis Using Present Worth Criterion, *The Engineering Economist*, Vol. 39, No. 2, Winter, pp. 113-138.

Choobineh, F. (1986), Justification of Flexible Manufacturing Systems, In: *Flexible Manufacturing Systems: Current Issues and Models*, Choobineh, F. (ed.), Industrial Engineering and Management Press, Atlanta, pp. 269-281.

Chung, K. (1993), The Contingent-Claims Approach to Investment Decisions, *Decision Sciences*, Vol. 24, No. 6, Nov./Dec., pp. 1215-1221.

Clark, J., Hindelang, T., and Pritchard, R. (1984), *Capital Budgeting: Planning and Control of Capital Expenditures*, Prentice-Hall, Inc., Englewood Cliffs, New Jersey.

Collier, P., and Gregory, A. (1995), Investment Appraisal in Service Industries: A Field Study Analysis of the U.K. Hotels Sector, *Management Accounting Research*, Vol. 6, pp. 33-57.

Cooley, J., and Hicks, J. Jr. (1983), A Fuzzy Set Approach to Aggregating Internal Control Judgments, *Management Science*, Vol. 29, No. 3, March, pp. 317-334.

Cramer, D. (1994), *Introducing Statistics for Social Research: Step-by-Step Calculations and Computer Techniques Using SPSS*, Routledge, London.

Creswell, J. (1994), *Research Design: Qualitative & Quantitative Approaches*, SAGE Publications, Inc., Thousand Oaks, California.

Currie, W. (1989), The Art of Justifying New Technology to Top Management, OMEGA, *International Journal of Management Science*, Vol. 17, No. 5, pp. 409-418.

Currie, W. (1994), *The Strategic Management of Advanced Manufacturing Technology*, The Chartered Institute of Management Accountants, London.

Datta, V., Sambasivarao, K., Kodali, R., and Deshmukh, S. (1992), Multi-attribute Decision Model Using the Analytic Hierarchy Process for the Justification of Manufacturing Systems, *International Journal of Production Economics*, Vol. 28, pp. 227-234.

Davis, E., and Pointon, J. (1994), *Finance and the Firm: An Introduction to Corporate Finance*, 2nd ed., Oxford University Press, Oxford.

Dhavale, D. (1989), Product Costing in Flexible Manufacturing Systems, *Journal of Management Accounting Research*, American Accounting Association, Vol. 1, Fall, pp. 66-88.

Drury, C., Braund, S., Osborne, P. and Tayles, M. (1993), *A Survey of Management Accounting Practices in UK Manufacturing Companies*, The Chartered Association of Certified Accountants, London, Chapter 6, pp. 41-48.

Dubois, D., and Prade, H. (1980), *Fuzzy Sets and Systems: Theory and Applications*, Academic Press, New York.

Dubois, D., and Prade, H. (1989), Fuzzy Sets, Probability and Measurement, *European Journal of Operational Research*, Vol. 40, No. 2, May 25, pp. 135-154.

Dugdale, D., and Abdel-Kader, M. (1998), Hierarchical Framework for Evaluating Investment Decisions in Advanced Manufacturing Technology: An Empirical Approach, *Journal of Applied Accounting Research*, Forthcoming.

Dugdale, D., and Jones, C. (1995), Financial Justification of Advanced Manufacturing Technology, In: *Issues in Management Accounting*, 2nd ed., Ashton, D., Hopper, T., and Scapens, R. (eds.), Prentice Hall, London, pp. 191-213.

Easterby-Smith, M., Thorpe, R., and Lowe, A. (1991), *Management Research: An Introduction*, SAGE Publications Ltd., London.

Elango, B., and Meinhart, W. (1994), Selecting a Flexible Manufacturing System - A Strategic Approach, *Long Range Planning*, Vol. 27, No. 3, June, pp. 118-126.

Farragher, E. (1986), Capital Budgeting Practices of Non-Industrial Firms, *The Engineering Economist*, Vol. 31, No. 4, Summer, pp. 293-302.

Finnie, J. (1988), The Role of Financial Appraisal in Decisions to Acquire Advanced Manufacturing Technology, *Accounting and Business Research,* Vol. 18, No. 70, pp. 133-139.

Fotsch, R. (1983), Machine Tool Justification Policies: Their Effect on Productivity and Profitability, *Journal of Manufacturing Systems,* Vol. 3, No. 2, pp. 169-195.

Frankfort-Nachmias, C., and Nachmias, D. (1992), *Research Methods in the Social Sciences,* Fourth Edition, Edward Arnold, London.

Franz, D., Duke, J., and Omer, K. (1995), Capital Budgeting: development of a Decision Approach for Determining Expected Value of Cash Flows, In: *Applications of Fuzzy Sets and the Theory of Evidence to Accounting,* Siegel, P., De Korvin, A., and Omer, K. (eds.), Jai Press Inc., London.

Fremgen, J. (1973), Capital Budgeting Practices: A survey, *Management Accounting (NY),* May, pp. 19-25.

FuziCalc (1994), *User's Guide* (version 1.5), FuziWare Inc.

Garrison, R. (1991), *Managerial Accounting: Concepts for Planning, Control, Decision Making,* 6th. ed., IRWIN, Boston.

Gerwin, D., and Kolodny, H. (1992), *Management of Advanced Manufacturing Technology: Strategy, Organization, and Innovation,* John Wiley & Sons, Inc., New York.

Ghauri, P., Gronhaug, K., and Kristianslund, I. (1995), *Research Methods in Business Studies: A Practical Guide,* Prentice Hall, New York.

Ghosh, B., and Wabalickis, R. (1991), A Comparative Analysis for the Justification of Future Manufacturing Systems, , *International Journal of Operations & Production Management,* Vol. 9, pp. 4-23.

Gibbons, J. (1993), *Nonparametric Statistics: An Introduction,* SAGE Publications, London.

Gluss, B. (1973), Fuzzy Multi-Stage Decision-Making, Fuzzy State and Terminal Regulators and Their Relationship to Non-fuzzy Quadratic State and Terminal Regulators, *International Journal of Control,* Vol. 17, No. 1, pp. 177-192.

Gowar, N., and Flegg, H. (1973), *Basic Mathematical Structures,* Vol. 1, Transworld Publishers Ltd., London.

Grinyer, J., and Green, C. (1997), *Payback and Economic Variables,* Paper Presented at the National Conference of The British Accounting Association, Birmingham, 24-26 March.

Groover, M., Weiss, M., Nagel, R., and Ordreg, N. (1986), *Industrial Robotics: Technology, Programming and Applications,* McGraw-Hill Book Co., New York.

Gutierrez, I., and Carmona, S. (1988), A Fuzzy Set Approach to Financial Ratio Analysis, *European Journal of Operational Research,* Vol. 36, No. 1, July, pp. 78-84.

Hamblin, D., and Lettman, A. (1996), Manufacturing Investment- Performance Causality in the UK clothing industry, *International Journal of Production Research,* Vol. 34, No. 9 (September), pp. 2421-2440.

Hammerbacher, I., and Yager, R. (1981), The Personalization of Security Selection: An Application of Fuzzy Set Theory, *Fuzzy Sets and Systems*, Vol. 5, 1981, pp. 1-9.

Harker, P. (1989), The Art Science of Decision Making: The Analytic Hierarchy Process, *The Analytic Hierarchy Process*, Golden, B. et al. (eds.), Springer- Verlag, New York, pp. 3-36.

Harper, R., Jr., Apostolou, N., and Hartman, B. (1992), The Analytical Hierarchy Process: An Empirical Examination of Aggregation and Hierarchical Structuring, *Behavioral Research in Accounting*, Vol. 4, pp. 96-112.

Harrison, M. (1990), *Advanced Manufacturing Technology Management*, Pitman Publishing, London.

Hendricks, J. (1988), Applying Cost Accounting to Factory automation, *Management accounting (USA)*, December, pp. 24-30.

Hilton, R. (1991), *Managerial Accounting*, McGraw-Hill, Inc., New York.

Hirst, I. (1988), *Business Investment Decisions*, Philip Allan, Oxford.

Ho, S., and Pike, R. (1991), Risk Analysis in Capital Budgeting Contexts: Simple or Sophisticated?, *Accounting and Business Research*, Vol. 21, No. 83, pp. 227-238.

Ho, S., and Pike, R. (1992), Adoption of Probabilistic Risk Analysis in Capital Budgeting and Corporate Investment, *Journal of Business Finance and Accounting*, Vol. 19, No. 3, April, pp. 387-405.

Hodder, J., and Riggs, H. (1985), Pitfalls in Evaluating Risky Projects, *Harvard Business Review*, Jan./Feb., pp. 128-135.

Hrbacek, K., and Jeck, T. (1984), *Introduction to Set Theory*, 2nd ed., Marcel Dekkee, Inc., New York.

Huang, P., Moore, L., and Shin, S. (1991), World-Class Manufacturing in the 1990s: Integrated TQC, JIT, FA, and TPM with worker Participation, *Manufacturing Review*, Vol. 4, No. 2, June, pp. 87-95.

Hundy B., and Hamblin, D. (1988), Risk and Assessment of Investment in New Technology, *International Journal of Production Research*, Vol. 26, No. 11, November, pp. 1799-1810.

Hussey, J., and Hussey, R. (1997), *Business Research: A Practical Guide for Undergraduate and Postgraduate Students*, Macmillan Press Ltd., Basingstoke, Great Britain.

Ijiri, Y. (1972), The Nature of Accounting Research, In: *Research Methodology in Accounting*, Sterling, R. (ed.), Scholars Book Co., pp. 59-69.

Institute of Production Engineers (1987), *Justifying Investment in Advanced Manufacturing Technology*, CIMA, I Prod E and Kogan Page Ltd.

Jain, R. (1976), Decision Making in the Presence of Fuzzy Variables, *IEEE Transactions on Systems, Man, and Cybernetics*, Vol. SMC-6, No. 10, October, pp. 698-703.

Jain, R. (1977), A Procedure for Multiple-aspects Decision Making Using Fuzzy Sets, *International Journal of Systems Science*, Vol. 8, No. 1, pp. 1-7.

Johnson, H., and Kaplan, R. (1991), *Relevance Lost: The Rise and Fall of Management Accounting*, Harvard Business School Press, Boston.

Kalyebara, B. (1996), *The Determination and Use of the Hurdle Rate in Capital Budgeting- Australian Experience*, Paper Presented at the National Conference of The British Accounting Association, Cardiff, 27-29 March.

Kandel, A. (1986), *Fuzzy Mathematical Techniques with Applications*, Addison-Wesley Publishing Company, Reading.

Kaplan, R. (1986), Must CIM be Justified by Faith Alone?, *Harvard Business Review*, March-April, pp. 87-93.

Kaplan, R., and Atkinson, A. (1989), *Advanced Management Accounting*, 2nd ed., Prentice-Hall, Englewood Cliffs, New Jersey, Chapter 12, pp. 473-521.

Kaufmann, A., and Gupta, M. (1985), *Introduction to Fuzzy Arithmetic: Theory and Applications*, Van Nostrand Reinhold Company, New York.

Kaufmann, A., and Gupta, M. (1991), *Fuzzy Mathematical Models in Engineering and Management Science*, Elsevier Science Publishing Company Inc., North-Holland.

Kee, R., and Bublitz, B. (1988), The Role of Payback in the Investment Process, *Accounting and Business Research*, Vol. 18, No. 70, pp. 149-155.

Kim, K., and Park, K. (1990), Ranking Fuzzy Numbers with index of optimism, *Fuzzy Sets and Systems*, Vol. 35, pp. 143-150.

Kim, S. (1982), An Empirical Study on the Relationship between Capital Budgeting Practices and Earnings Performance, *The engineering Economist*, Vol. 27, No. 3, pp. 185-196.

Kim, S., Crick, T., and Kim, Seung H. (1986), Do Executives Practice What Academics Preach?, *Management Accounting (USA)*, November, pp. 49-52.

Klahorst, T. (1984), FMS, *Flexible Manufacturing Systems*, Holland, J. (ed.), Society of Manufacturing Engineering, Michigan, pp. 3-6.

Klahorst, H. (1986), How to Justify Multimachine Systems, *Justifying New Manufacturing Technology*, Meredith, J. (ed.), Industrial Engineering and Management Press, Atlanta, pp. 149-152.

Klammer, T. (1970), Empirical Evidence of the Adoption of sophisticated capital Budgeting Techniques, *Journal of Business*, July, pp. 387-397.

Klammer, T. (1973), The Association of Capital Budgeting Techniques with Firm Performance, *The Accounting Review*, April, pp. 353-364.

Klammer, T., and Walker, M. (1984), The Continuing Increase in the Use of Sophisticated Capital Budgeting Techniques, *California Management Review*, Vol. 27, No. 1, Fall, pp. 137-148.

Klammer, T., Koch, B., and Wilner, N. (1991), Capital Budgeting Practices- A Survey of Corporate Use, *Journal of Management Accounting Research*, Vol. 3, Fall, pp. 113-130.

Klee, A. (1971), The Role of Decision Models in the Evaluation of Competing Environmental Health Alternatives, Management Science, Vol. 18, No. 2, October, pp. B-52- B-67.

Klir, G., and Floger, T. (1988), *Fuzzy Sets, Uncertainty, and Information*, Prentice-Hall International, Inc., USA.

Knott, K., and Getto, R. (1982), A Model for Evaluating Alternative Robot Systems Under Uncertainty, *International Journal of Production Research*, Vol. 20, pp. 155-165.

Kochan, A., and Cowan, D. (1986), *Implementing CAM: Computer Integrated Manufacturing*, IFS (Publications) LTD, Bedford, UK.

Koening, D. (1990), *Computer Integrated Manufacturing: Theory and Practice*, Hemisphere Publishing Corporation, New York.

Kolli, S., Wilhelm, M., Parsaei, H., and Liles, D. (1992), A Classification Scheme for Traditional and Non-traditional Approaches to Economic Justification of Advanced Automated Manufacturing Systems, In: *Economic and Financial Justification of Advanced Manufacturing Technologies*, Parsaei, H., Sullivan, W., and Hanley, T. (eds.), Elsevier Science Publishers, pp. 165-187.

Kruse, R., Gebhardt, J., and Klawonn, F. (1994), *Foundations of Fuzzy Systems*, John Wiley & Sons, Chichester.

Kulatilaka, N. (1984), Financial, Economic and Strategic Issues Concerning the Decision to Invest in Advanced Automation, *International Journal of Production Research*, Vol. 22, No.6, pp. 949-968.

Kusiak, A. (1985), Flexible Manufacturing Systems: A Structural Approach, *International Journal of Production Research*, Vol. 23, No. 6, Nov./Dec., pp.1057-1073.

Kvanli, A., Guynes, C., and Pavur, R. (1992), Introduction to Business statistics: A Computer Integrated Approach, West Publishing Co., New York.

Laarhoven, P., and Pedrgcz, W. (1983), A Fuzzy Extension of Saaty's Priority Theory, *Fuzzy Sets and Systems*, Vol. 11, No. 3, pp 229-241.

Lavelle, J., and Liggett, H. (1992), Economic Methods for Evaluating Investments in Advanced Manufacturing Technologies, In: *Economic and financial Justification of Advanced Manufacturing Technologies*, Parsaei, H., Sullivan, W., and Hanley, T. (eds.), Elsevier Science Publishers, pp. 119-139.

Lefley, F. (1994a), Capital Investment Appraisal of Advanced Manufacturing Technology, *International Journal of Production research*, Vol. 32, No. 12, pp. 2751-2776.

Lefley, F. (1994b), AMT Project Hurdle Rates too High?, *Management Accounting (UK)*, May, pp. 37, 46.

Lefley, F. (1996), Strategic Methodologies of Investment Appraisal of AMT Projects: A Review and Synthesis, *The Engineering Economist*, Vol. 41, No. 4, Summer, pp. 345-363.

Lefley, F. (1997), Modified Internal Rate of Return: Will It Replace IRR?, *Management Accounting*, January, pp. 64-65.

Liggett, H., Trevino, J., and Lavelle, J. (1992), Activity-Based Cost Management Systems in Advanced Manufacturing Environment, In: *Economic and financial Justification of Advanced Manufacturing Technologies*, Parsaei, H., Sullivan, W., and Hanley, T. (eds.), Elsevier Science Publishers, pp. 39-55.

Lin, T., and O'leary, D. (1989), Cost Variance Investigation with Fuzzy Sets, In: *Applications of Fuzzy Set Methodologies in Industrial Engineering*, Evans, G., Karwowski, W., and Wilhelm, M. (eds.), Elsevier, Amsterdam, pp. 213-222.

Liu, C., and Wang, M. (1994), A Multiple Criteria Linguistic Decision Model (MCLDM) for Human Decision Making, *European Journal of Operational Research*, Vol. 76, pp. 466-485.

Luckett, P. (1984), ARR vs. IRR: A Review and Analysis, *Journal of Business and Accounting*, Vol. 11, No. 2, Summer, pp. 213-231.

Lumby, S. (1995), *Investment Appraisal and Financial Decisions*, 5th ed., Chapman & Hall, London.

Medearis, H., Helms, M., and Ettkin, L. (1990), Justifying Flexible Manufacturing Systems (FMS) from a Strategic Perspective, *Manufacturing Review*, Vol. 3, No. 4, December, pp. 219-223.

Mensah, Y., and Miranti, P. (1989), Capital Expenditure Analysis and Automated Manufacturing Systems: A Review and Synthesis, *Journal of Accounting Literature*, Vol. 8, pp. 181-207.

Meredith, J. (1989) Management survey on Justifying New Manufacturing Technologies, *International Journal of Computer Integrated Manufacturing*, Vol. 2, No. 3, pp. 164-170.

Meredith, J., and Hill, M. (1987), Justifying New Manufacturing Systems: A Managerial Approach, *Sloan Management Review*, summer, pp. 49-61.

Meredith, J., and McCutcheon, D. (1991), Recent Advances in Justifying CIM Technologies, *Computer-Integrated Manufacturing Systems*, Vol. 4, No. 4, November, pp. 248-253.

Meredith, J., and Suresh, N. (1986), Justification Techniques for Advanced Manufacturing Technologies, *International Journal of Production Research*, Vol. 24, No. 5, Sept./Oct., pp. 1043-1057.

Merrett, A., and Sykes, A. (1980), *The Finance and Analysis of Capital Projects*, 2nd. ed., Longman, London.

Miltenburg, G., and Krinsky, I. (1987), Evaluating Flexible Manufacturing, *IIE Transactions*, Vol. 19, No. 2, June, pp. 222-233.

Mintzberg, H. (1994), The Rise and Fall of strategic Planning, Prentice Hall, London.

Mintzberg, H., Quinn, J., and Ghoshal, S. (1995), The strategy Process, Prentice Hall, London.

Mohanty, R., and Venkatarman, S. (1993), Use of the Analytic Hierarchy Process for Selecting Automated Manufacturing Systems, *International Journal of Operations & Production Management*, Vol. 13, No. 8, pp. 45-57.

Nagarur, N. (1992), Some Performance Measures of Flexible Manufacturing Systems, *International Journal of Production Research*, Vol. 30, No. 4, April, pp. 799-809.

Naik, B., and Chakravarty, A. (1992), Strategic Acquisition of New Manufacturing Technology, *International Journal of Production Research*, Vol. 30, No. 7, July, pp. 1575-1601.

Nelson, C. (1986), A Scoring Model for Flexible Manufacturing Systems Project Selection, *European Journal of Operational Research*, Vol. 24, pp. 346-359

Newell, R. (1993), Questionnaires, In: Researching Social Life, Gilbert, N. (ed.), SAGE Publications, London, pp. 94-115.

Nixon, B. (1995), Technology investment and management accounting practice, *British Journal of Management*, Vol. 6, pp. 271-288.

Norusis, M. (No date), SPSS 6.1: Guide to Data Analysis, Prentice Hall, Inc., New Jersey.

O'Brien, M. (1995), Risk Assessment, *Journal of the Institute of Chartered Accountants in Ireland*, October, pp. 18-19.

O'Brien, C., and Smith, S. (1993), Design of the Decision Process for Strategic Investment in Advanced Manufacturing Systems, *International Journal of Production Economics*, Vol. 30-31, pp. 309-322.

Okuda, T., Tanaka, H., and Asai, K. (1978), A Formulations of Fuzzy Decision Problems with Fuzzy Information Using Probability Measures of Fuzzy Events, *Information and Control*, Vol. 38, No. 2, August, pp. 135-147.

Pal, S., and Dutta Majumder, D. (1987), *Fuzzy Mathematical Approach to Pattern Recognition*, Wiley Eastern Limited, New Delhi.

Park, C., and Son, Y. (1988), An Economic Evaluation Model for Advanced Manufacturing Systems, *The Engineering Economist*, Vol. 34, No. 1, Fall, pp. 1-26.

Parsaei, H., and Wilhelm, M. (1989), A Justification Methodology for Automated Manufacturing Technologies, *Computers and Industrial Engineering*, Vol. 16, No. 3, pp. 363-373.

Pike, R. (1982), *Capital Budgeting in the 1980s: A major Survey of the Investment Practices in Large Companies*, The Institute of Cost and Management Accountants, London.

Pike, R. (1983), A Review of Recent Trends in Formal Capital Budgeting Processes, *Accounting and Business Research*, Summer, pp. 201-208.

Pike, R. (1985), Owner-Manager Conflict and the Role of the Payback Method, *Accounting and Business research*, Winter, pp. 47-51.

Pike, R. (1988a), An Empirical Study of the Adoption of Sophisticated Capital Budgeting Practices and Decision-Making effectiveness, *Accounting and Business Research*, Vol. 18, No. 72, pp. 341-351.

Pike, R. (1988b), The Capital Budgeting Revolution, *Management Accounting (UK)*, October, pp. 28-30.

Pike, R. (1989), Do Sophisticated Capital Budgeting Approaches Improve Investment Decision-Making Effectiveness?, *The Engineering Economist*, Vol. 34, No. 2, Winter, pp. 149-161.

Pike, R. (1996), A Longitudinal Survey on Capital Budgeting Practices, *Journal of Business Finance & Accounting*, Vol. 23, No. 1, January, pp. 79-92.

Pike, R., and Dobbins, R. (1986), *Investment Decisions and Financial Strategy*, Philip Allan, Oxford.

Pike, R., and Roberts, B. (1980), Investment Hurdle Rates and the Cost of Capital, *Management Accounting (UK)*, May, pp. 26-28.

Pike, R., and Wolfe, M. (1988), *Capital Budgeting for the 1990's*, The Chartered Institute of Management Accountants (CIMA), London.

Pike, R., Sharp, J., and Price, D. (1989), AMT Investment in the Larger UK Firm, *International Journal of Operations and Production Management*, Vol. 9, No. 2, pp. 13-26.

Pohjola, V., and Turunen, I. (1990), Estimating the internal Rate of Return from Fuzzy Data, *Engineering Costs and Production Economics*, Vol. 18, pp. 215-221.

Porter, M. (1980), Competitive Strategy, Collier Macmillan, New York.

Primrose, P. (1988), AMT Investment and Costing Systems, *Management Accounting*, October, pp. 26-27.

Primrose, P. (1991), *Investment in Manufacturing Technology*, Chapman & Hall London.

Primrose, P. (1992a), R & D Investment - Can It Be Justified?, *Professional Engineering*, March, pp. 19-20.

Primrose, P. (1992b), Is Anything Really Wrong With Cost Management?, *Cost Management*, Spring, pp. 48-57.

Primrose, P., and Leonard, R. (1986a), Evaluating the 'Intangible' Benefits of Flexible Manufacturing Systems by Use of Discounted Cash Flow Algorithms Within a Comprehensive Computer Program, In: *Flexible Manufacturing Systems: Current Issues and Models*, Choobineh, F., and Suri, R. (eds.), Industrial Engineering and Management Press , Atlanta, pp. 289-294.

Primrose, P., and Leonard, R. (1986b), The Use of a Conceptual Model to Evaluate Financially Flexible Manufacturing System Projects, In: *Flexible Manufacturing Systems: Current Issues and Models*, Choobineh, F., and Suri, R. (eds.), Industrial Engineering and Management Press , Atlanta, pp. 282-288.

Primrose, P., and Leonard, R. (1987), Automation Needs No Justification, *New Scientist*, Vol. 115, No. 1576, 3 September, pp. 60-62.

Primrose, P., and Leonard, R. (1988), Investing to Improve Product Quality, *IJQRM*, Vol. 5, No. 4, pp. 38-45.

Primrose, P., Creamer, G., and Leonard, R. (1985), Identifying and Quantifying the Company-Wide Benefits of CAD Within the Structure of a Comprehensive Investment Program, *Computer Aided Design*, Vol. 17, No. 1, Jan./Feb., pp. 3-8.

Putrus, R. (1990), Accounting for Intangible in Integrated Manufacturing, *Information Strategy: The Executives Journal*, Vol. 6, No. 4, Summer, pp. 25-30.

QSR NUD.IST (1995), *User's Guide*, Qualitative Solutions and Research Pty Ltd.

Randhawa, S., and West, T. (1992), Evaluating Automated Manufacturing Technologies: Part I- Concepts and Literature Review, *Computer- Integrated Manufacturing Systems*, Vol. 5, No. 3, August, pp. 208-218.

Rangone, A. (1997), Linking Organizational Effectiveness, Key Success Factors and Performance Measures: An Analytical Framework, *Management Accounting Research*, Vol. 8, June, pp. 207-219.

Rayburn, L. (1989), *Principles of Cost Accounting Using a Cost Management Approach*, 4th. ed., IRWIN, Boston.

Reekie, W., and Crook, J. (1995), *Managerial Economics: A European Text*, 4th ed., Prentice-Hall, New York.

Riggs, J., and West, T. (1986), *Engineering Economics*, McGraw-Hill Book Company, New York.

Ronen, J., and Sorter, G. (1972), Relevant Accounting, *Journal of Business*, Vol. 45, April, pp. 258-282.

Roy, U., and Slocum, D. (1992), Cost Accounting and the Justification of Advanced Manufacturing Technologies, In: *Economic and Financial Justification of Advanced Manufacturing Technologies*, Parsaei, H., Sullivan, W., and Hanley, T. (eds.), Elsevier Science Publishers, pp. 57-67.

Rueff, M., and Jeger, M. (1970), *Sets and Boolean Algebra*, George Allen and Unwin Ltd., London.

Ryan, B., Scapens, R., and Theobald, M. (1992), *Research Method and Methodology in Finance and Accounting*, Academic Press Limited, London.

Saaty, T. (1980), *The Analytic Hierarchy Process*, McGraw Hill, New York.

Saaty, T. (1990), *The Analytic Hierarchy Process: Priority Setting, Resource Allocation (Multicriteria Decision Making)*, Vol. 1, RWS Publications, Pittsburgh, USA.

Samuels, J., and Wilkes, F. (1986), *Management of Company Finance*, 4th ed., Van Nostrand Reinhold (UK) Co. Ltd.

Sangster, A., and Baldwin-Morgan, A. (1996), Field-Based Research in Accounting Information Systems, *Informal proceedings of Beneath the Numbers: Reflections on the Use of Qualitative Methods in Accounting Research*, Portsmouth Business School, 4-5 January, pp. 508-530.

Sarantakos, S. (1993), *Social Research*, The Macmillan Press Ltd., Basingstoke, Great Britain.

Scapens, R., Turley, S., Burns, J., Lewis, L., Joseph, N., and Southworth, A. (1996), *External Reporting and Management Decisions: A Study of Their Interrelationship in UK Companies*, Chartered Institute of Management Accountants, London.

Schall, L., Sundem, G., and Geijsbeek, W., Jr. (1978), Survey and analysis of Capital Budgeting Methods, *The Journal of Finance*, March, pp. 281-287.

Schoemaker, P. (1979), The Role of Statistical Knowledge in Gambling Decisions: Moment vs Risk Dimension Approaches, *Organizational Behavior and Human Performance*, Vol. 24, August, pp. 1-17.

Schroeder, R. (1989), *Operations Management: Decision Making in the Operations Function*, 3rd ed., McGraw-Hill Book Co., London.

Scott, D. (1990), *Understanding and Managing Investment Risk & Return*, McGraw-Hill Book Co., London.

Shank, J. (1996), Analysing Technology Investments-from NPV to Strategic Cost Management (SCM), *Management Accounting Research*, Vol. 7, No. 2, June, pp. 185-197.

Shao, L., and Shao, A. (1996), Risk Analysis and Capital Budgeting Techniques of U.S. Multinational Enterprises, *Managerial Finance*, Vol. 22, No. 1, pp. 41-57.

Sheehan, M. (1994), The Investment Decision-Making Process and Investment Determinants in the Northern Ireland Manufacturing Sector, *Regional Studies*, Vol. 28.5, pp. 511-520.

Siegel, P., De Korvin, A., and Omer, K. (eds.) (1995), *Applications of Fuzzy Sets and the Theory of Evidence to Accounting*, Jai Press Inc., London.

Siegel, S., and Castellan, N., Jr. (1988), *Nonparametric Statistics for the Behavioral Sciences*, 2nd edition, McGraw-Hill Book Co., New York.

Sirkin, R. M. (1995), *Statistics for the Social Sciences*, SAGE Publications, London.

Slagmulder, R. (1997), Using Management Control Systems to Achieve Alignment between Strategic Investment Decisions and strategy, *Management Accounting Research*, Vol. 8, pp. 103-139.

Slagmulder, R., and Bruggeman, W. (1992), Justification of strategic investment in flexible manufacturing technology, *Integrated Manufacturing Systems*, Vol. 3, pp. 4-14.

Slagmulder, R., Bruggeman, W., and Wassenhove, L. (1995), An Empirical Study of Capital Budgeting Practices for Strategic Investments in CIM Technologies, *International Journal of Production Economics*, Vol. 40, pp. 121-152.

Smithson, M. (1987), *Fuzzy Set Analysis for Behavioural and Social Sciences*, Springer-Verlag, New York.

Somers, T., and Gupta, Y. (1991), Factory Automation: U. S. Versus Japanese Experiences, *Manufacturing Review*, Vol. 4, No. 3, September, pp. 157-167.

Son, Y. (1991a), A Cost Estimation Model for Advanced Manufacturing Systems, *International Journal of Production Research*, Vol. 29, No. 3, March, pp. 441-452.

Son, Y. (1991b), A Framework for Modern Manufacturing Economics, *International Journal of Production Research*, Vol. 29, No. 12, Dec., pp. 2483-2499.

Soni, R., Parsaei, H., and Khosravi-Kamrani, A. (1990), Economics Justification Methods for Advanced Manufacturing Systems, *Proceedings of IIE Conference*, May, pp. 146-151.

Soni, R., Parsaei, H., and Liles, D. (1992), Economic and Financial Justification Methods for Advanced Automated Manufacturing: An Overview, In: *Economics of Advanced Manufacturing Systems*, Parsaei, H., and Mital, M. (eds.), Chapman & Hall, London, pp. 3-19.

Srinivasan, V., and Millen, R. (1986), Evaluating Flexible Manufacturing Systems as a Strategic Investment, *Proceedings of the Second ORSA/TIMS Conference on Flexible Manufacturing Systems: Operations Research Models and Applications*, Stecke, K., and Suri, R. (eds.), Elsevier Science Publishers, Amsterdam, pp. 83- 93.

Stevens, G., Jr. (1979), *Economic and Financial Analysis of Capital Investments*, John Wiley & Sons, New York.

Suresh, N., and Meredith, J. (1985), Justifying Multimachine systems: An Integrated Strategic Approach, *Journal of Manufacturing Systems*, Vol. 4, No. 2, pp. 117-134.

Swamidass, P., and Waller, M. (1990), A Classification of Approaches to Planning and Justifying New Manufacturing Technologies, *Journal of Manufacturing Systems*, Vol. 9, No. 3, pp. 181-193.

Swann, K., and O'Keefe, W. (1990), Advanced Manufacturing Technology: Investment Decision Process. Part 1, *Management Decision*, Vol. 28, No. 1, pp. 20-31.

Tannous, G. (1996), Capital Budgeting for Volume Flexible Equipments, *Decision Science*, Vol. 27, No. 2, Spring, pp. 157-184.

Tombari, H. (1986), To Buy or Not to Buy?, In: *Justifying New Manufacturing Technology*, Meredith, J. (ed.), Industrial Engineering and Management Press, Atlanta, pp. 107-110.

Tomkins, C. (1991), *Corporate Resource Allocation; Financial, strategic and Organizational Perspectives*, Basil Blackwell, Oxford.

Tomkins, C., and Carr, C. (1996), Reflections on the Papers in this Issue and a Commentary on the State of Strategic Management Accounting, *Management Accounting Research*, Vol. 7, No. 2, June, pp. 271-280.

Van Cauwenbergh, A., Durinck, E., Martens, R., Laveren, E., and Bogaert, I. (1996), On the Role and Function of Formal Analysis in Strategic Investment Decision Processes: Results from an Empirical Study in Belgium, *Management Accounting Research*, Vol. 7, No. 2, June, pp. 169-184.

Wabalickis, R. (1988), Justification of FMS with the Analytic Hierarchy, *Journal of Manufacturing Systems*, Vol. 7, No. 3, pp. 175-182.

Wang, J., Liittschwager, J., and Raz, T. (1989), Equipment Life Estimation Using Fuzzy Set Theory, In: *Applications of Fuzzy Set Methodologies in Industrial Engineering*, Evans, G., Karwowski, W., and Wilhelm, M. (eds.), Elsevier, Amsterdam, pp. 197-211.

Wang, M., and Liang, G. (1995), Benefit/Cost Analysis Using Fuzzy Concept, *The Engineering Economist*, Vol. 40, No. 4, Summer, pp. 359-376.

Ward, T. (1985), Discounted Fuzzy Cash Flow Analysis, *Proceedings of Institute of Industrial Engineers*, December, pp. 476-481.

Ward, T. (1989), *Fuzzy Discounted Cash Flow Analysis*, In: *Applications of Fuzzy Set Methodologies in Industrial Engineering*, Evans, G., Karwowski, W., and Wilhelm, M. (eds.), Elsevier, Amsterdam, pp. 91-102.

Weingartner, H. (1969), Some New Views on the Payback Period and Capital Budgeting Decisions, *Management Science*, Vol. 15, No. 12, August, pp. 594-607.

Wilhelm, M., and Parsaei, H. (1988), A Model for Selecting and Evaluating Flexible Automation Systems, *Proceedings of the Institute of Industrial Engineers Integrated Systems Conference*, pp. 341-346.

Wilhelm, M., and Parsaei, H. (1991), A Fuzzy Linguistic Approach to Implementing a Strategy for Computer Integrated Manufacturing, *Fuzzy Sets and Systems*, Vol. 42, pp. 191-204.

Wilner, N., Koch, B., and Klammer, T. (1992), Justification of High Technology Capital Investment- An Empirical Study, *The Engineering Economist*, Vol. 37, No. 4, Summer, pp. 341-353.

Wolper, J. (1987), Conference Paper: A New Type of FMS: its Design, its Efficiency, its Feasibility, *International Journal of Production Research*, Vol. 25, No. 11, Nov., pp. 1611-1624.

Woods, M., Pokorny, M., Lintner, V., and Blinkhorn, M. (1985), Appraising Investment in New Technology: The Approach in Practice, *Management Accounting (UK)*, October, pp. 42-43.

Yin, R. (1989), *Case Study Research: Design and Methods*, SAGE Publications, Inc., Newbury Park, California.

Zadeh, L. (1965), Fuzzy Sets, *Information and Control*, Vol. 8, pp. 338-353.

Zadeh, L. (1968), Probability Measures of Fuzzy Events, *Journal of Mathematical Analysis and Applications*, Vol. 23, No. 2, August, pp. 421-427.

Zadeh, L. (1972), A Fuzzy Set Theoretic Interpretation of Linguistic Hedges, *Journal of Cybernetics*, Vol. 2, No. 3, pp. 4-34.

Zadeh, L. (1973), Outline of a New Approach to the analysis of Complex Systems and Decisions Processes, *IEEE Transactions of Systems, Man, and Cybernetics*, Vol. SMC-3, No. 1, January, pp. 28-44.

Zadeh, L. (1975a), The Concept of a Linguistic Variable and its application to Approximate Reasoning - I, *Information Science*, Vol. 8, pp. 199-249.

Zadeh, L. (1975b), The Concept of a Linguistic Variable and its application to Approximate Reasoning - II, *Information Science*, Vol. 8, pp. 301-357.

Zadeh, L. (1975c), The Concept of a Linguistic Variable and its application to Approximate Reasoning - III, *Information Science*, Vol. 9, pp. 43-80.

Zadeh, L., Fu, K., Tanaka, K., and Shimura, M. (1975), *Fuzzy Sets and Their Applications to Cognitive and Decision Processes*, Academic Press, Inc., New York.

Zebda, A. (1984), The Investigation of Cost Variances: A Fuzzy Set Theory Approach, *Decision Sciences*, Vol. 15, No. 3, Summer, pp. 359-388.

Zebda, A. (1989), Fuzzy Set Theory and Accounting, *Journal of Accounting Literature*, Vol. 8, pp. 76-105.

Zebda, A. (1991), The problem of Ambiguity and Vagueness in Accounting, *Behavioural Research in Accounting*, Vol. 3, 1991, pp. 117-145.

Zimmermann, H. (1987), *Fuzzy Sets, Decision Making and Expert Systems*, Kluwer Academic Publishers, Boston.

Zimmermann, H. (1991), *Fuzzy Set Theory and its Applications*, 2nd ed., Kluwer Academic Publishers, Boston.

Appendices

Appendix 1 - The Preliminary Letter to the Sample Companies

25 November, 1995

Dear Sir:

I am conducting a survey on behalf of Bristol Business School to explore current methods used in evaluating investment decisions regarding the adoption of advanced manufacturing technology and to identify major factors affecting these decisions in practice.

Advanced Manufacturing Technology (AMT) is used to refer to any computerised manufacturing process form such as Computer Numerically Controlled machines (CNC), robots, Computer Aided Design (CAD), Flexible Manufacturing System (FMS), etc.

I would be very grateful if you could participate in this survey. If so, I will send you a questionnaire which will take 10-15 minutes to complete. Your response will be treated with perfect confidentiality and used only for research purposes. If you participate you will be sent a copy of the results of the survey, and given the opportunity to keep in touch with the research.

It would also be helpful if you could answer the four questions below and return this letter in the enclosed pre-paid envelope:

1. Are you or a colleague willing to participate in the survey? □Yes □No

State name and address if different to your own: ....
....
....
....

2. Has your firm invested in advanced manufacturing technology (AMT)?
□Yes □No

3. Have you ever been involved in the evaluation of investment decisions?
   a. AMT investment proposal ☐ Yes ☐ No
   b. Other investment proposal ☐ Yes ☐ No

4. Could you rank the following considerations (1 most important, 3 least important) as regards to both AMT and other investments:

	AMT investments	Other investments
the financial payoff	____	____
non-financial benefits	____	____
risk	____	____

I appreciate your help in this study and look forward to receiving your reply.

Yours sincerely,

Magdy Abdel-Kader
Bristol Business School

Appendix 2 - The Cover Letter of the Questionnaire

Cover letter (A)

January 1996

Dear Sir:

Thank you very much for your response to my letter. Your participation in the survey will be of great help to the research. The aim of this survey is to explore current methods used in evaluating investment decisions regarding the adoption of advanced manufacturing technologies and to identify major factors affecting these decisions in practice.

The questionnaire will take 10-15 minutes to complete. The information you can provide is very important as it will reflect the realities of investment decision-making.

Your response will be treated with complete confidentiality and used only for research purposes. You will be sent a copy of the results of the survey, and given the opportunity to keep in touch with the research.

After analysing the questionnaire results, I would like to undertake a limited number of interviews with practitioners. It would be extremely helpful if you could spare some time to participate in these interviews. Interviews might take about an hour and would be arranged at your convenience. If you are willing to participate in such interviews, please indicate on the last page of the questionnaire.

Please return the completed questionnaire in the enclosed pre-paid envelope. I would very much appreciate your willingness to help in this study and look forward to receiving your reply.

Yours sincerely,

Magdy Abdel-Kader
Bristol Business School

Cover letter (B)

January 1996

Dear Sir:

I am conducting a survey on behalf of Bristol Business School. The aim of this survey is to explore current methods used in evaluating investment decisions regarding the adoption of advanced manufacturing technologies and to identify major factors affecting these decisions in practice.

I would be very grateful if you could participate in this questionnaire. It will take 10-15 minutes to complete. The information you can provide is very important as it will reflect the realities of investment decision-making.

Your response will be treated with complete confidentiality and used only for research purposes. If you participate you will be sent a copy of the results of the survey, and given the opportunity to keep in touch with the research.

After analysing the questionnaire results, I would like to undertake a limited number of interviews with practitioners. It would be extremely helpful if you could spare some time to participate in these interviews. Interviews might take about an hour and would be arranged at your convenience. If you are willing to participate in such interviews, please indicate on the last page of the questionnaire.

Please return the completed questionnaire in the enclosed pre-paid envelope. I would very much appreciate your willingness to help in this study and look forward to receiving your reply.

Yours sincerely,

Magdy Abdel-Kader
Bristol Business School

Appendix 3 - The Questionnaire

Investments in AMT

In this questionnaire AMT (Advanced Manufacturing Technology) is used to refer to any computerised manufacturing process form such as Computer Numerically Controlled machines (CNC), robots, Computer Aided Design (CAD), Flexible Manufacturing System (FMS), etc.

Q1 Have you ever been involved in the evaluation of investment decisions (whether approved or not)

a. AMT investment proposal	1□ Yes	2□ No
b. Other investment proposal	1□ Yes	2□ No

Q2 If your answer to (Q1.a) was YES, how would you rate the following in evaluating an AMT investment proposal:

	Very Easy				Very Difficult
Quantifying intangible factors	1	2	3	4	5
Predicting cash flows	1	2	3	4	5
Estimating labour savings	1	2	3	4	5
Estimating material savings	1	2	3	4	5
Estimating energy savings	1	2	3	4	5
Assessing risk	1	2	3	4	5
Estimating project life	1	2	3	4	5
Other (please specify)					
______________	1	2	3	4	5
______________	1	2	3	4	5
______________	1	2	3	4	5

Q3 Has your firm invested in AMT? 1□ Yes 2□ No

(If NO go to Question 6)

Q4 If your answer to (Q3) was Yes, please indicate which type(s) of AMT has your firm invested in by ticking the relevant box:

	(1) Pre 1985	(2) 1985-90	(3) 1991-96
Computer Numerical Control (CNC)	☐	☐	☐
Robotics	☐	☐	☐
Automated Material Handling (AMH)	☐	☐	☐
Flexible Manufacturing Systems(FMS)	☐	☐	☐
Computer Aided Design (CAD)	☐	☐	☐
Computer Aided Manufacturing (CAM)	☐	☐	☐
Computer Integrated Manufacturing (CIM)	☐	☐	☐
Other (please specify)			
________________	☐	☐	☐
________________	☐	☐	☐

Q5 If your answer to (Q3) was Yes, please indicate the life expected for AMT proposals:

	(1) 3 years or less	(2) 4-6 years	(3) 7-10 years	(4) more than 10 years
Computer Numerical Control (CNC)	☐	☐	☐	☐
Robotics	☐	☐	☐	☐
Automated Material Handling (AMH)	☐	☐	☐	☐
Flexible Manufacturing Systems(FMS)	☐	☐	☐	☐
Computer Aided Design (CAD)	☐	☐	☐	☐
Computer Aided Manufacturing (CAM)	☐	☐	☐	☐
Computer Integrated Manufacturing (CIM)	☐	☐	☐	☐
Other (please specify)				
________________	☐	☐	☐	☐
________________	☐	☐	☐	☐

Q6 You may have used one or more of the following techniques or criteria in evaluating investment proposals. Please indicate their relative importance by circling the appropriate number.

	Not Important				Extremely Important
a. Financial Appraisal techniques:					
Return on Investment (ROI)	1	2	3	4	5
Payback (PB)	1	2	3	4	5
Discounted Payback	1	2	3	4	5
Net Present Value (NPV)	1	2	3	4	5
Internal Rate of Return (IRR)	1	2	3	4	5
Other (please specify)					
________________	1	2	3	4	5
________________	1	2	3	4	5
________________	1	2	3	4	5
b. Non Financial Criteria:					
Consistency with corporate strategy	1	2	3	4	5
Improved company image	1	2	3	4	5
Requirements of customers	1	2	3	4	5
Keeping up with competition	1	2	3	4	5
Obtaining greater manufacturing flexibility	1	2	3	4	5
The ability to expand in the future	1	2	3	4	5
Quality and reliability of outputs	1	2	3	4	5
Employee morale	1	2	3	4	5
Experience with new technology	1	2	3	4	5
Reduced lead-times	1	2	3	4	5
Reduced inventory levels	1	2	3	4	5
Other (please specify)					
________________	1	2	3	4	5
________________	1	2	3	4	5
________________	1	2	3	4	5

	Not Important				Extremely Important
c. Risk analysis techniques:					
Adjust required payback period	1	2	3	4	5
Adjust required return on investment	1	2	3	4	5
Adjust discount rate	1	2	3	4	5
Use of conservative cash flow forecasts	1	2	3	4	5
Sensitivity analysis	1	2	3	4	5
Computer simulation	1	2	3	4	5
Probability analysis	1	2	3	4	5
Capital asset pricing model (Beta analysis)	1	2	3	4	5
Other (please specify)					
________________	1	2	3	4	5
________________	1	2	3	4	5
________________	1	2	3	4	5
d. Other financial criteria:					
Possibility of massive loss	1	2	3	4	5
Variability of project outcome	1	2	3	4	5
Impact on company liquidity	1	2	3	4	5
Other (please specify)					
________________	1	2	3	4	5
________________	1	2	3	4	5
________________	1	2	3	4	5

Q7 If your firm uses the payback period method to evaluate an investment proposal, please indicate which of the following best reflects the required payback period which is most frequently used by ticking <u>only one box</u>

Less than 1 year	1☐
1 - 3 years	2☐
4 - 5 years	3☐
More than 5 years	4☐

Q8 If your firm uses discounting methods to evaluate investment proposals, please indicate which of the following best reflects the range of minimum rates of return or discount rates required by ticking **only one box**

Less than 10%	1☐
10 - 15%	2☐
16 - 20%	3☐
21- 25%	4☐
More than 25%	5☐

Q9 If your firm uses more than one financial appraisal technique to evaluate investment proposals, please indicate which of the following represents the most important technique by ticking **only one box**:

1. Return on Investment (ROI)	☐
2. Payback (PB)	☐
3. Discounted Payback	☐
4. Net Present Value (NPV)	☐
5. Internal Rate of Return (IRR)	☐
6. Other (please specify)	
________________	☐
________________	☐

Q10 Please indicate how you consider the following benefits of AMT investment proposals by circling **only 1,2 or 3** in the relevant line:

1 ⟶ if you consider this benefit mainly in financial terms,
2 ⟶ if you consider this benefit mainly non-financially, or
3 ⟶ if you don't consider this benefit at all.

Reduced labour costs	1	2	3
Reduced material costs	1	2	3
Reduced scrap and rework costs	1	2	3
Reduced inventories level	1	2	3
Reduced lead times	1	2	3
Reduced after sale costs such as warranty etc.	1	2	3
Savings from less frequent set-ups	1	2	3

Improved product quality	1	2	3
Retention of market share	1	2	3
Increased market share	1	2	3
Faster response to market needs	1	2	3
Improved company image	1	2	3
Improved competitive position	1	2	3
Increased manufacturing capacity	1	2	3
Easier production scheduling	1	2	3
Floor space reduction	1	2	3
Consistency with corporate strategy	1	2	3
Obtaining experience of new technology	1	2	3
Effects on employee morale	1	2	3
Greater manufacturing flexibility	1	2	3
Increased sales volume	1	2	3
Others (please specify)			
________________	1	2	3
________________	1	2	3

Q11 Please indicate whether you agree or disagree with the following statements by circling the appropriate number.

	Strongly disagree				Strongly agree
a. It is difficult to get AMT investment proposals approved because of stringent financial criteria.	1	2	3	4	5
b. Some of potential benefits of AMT are not taken into account because they are difficult to express in financial terms.	1	2	3	4	5
c. It is possible to implement an AMT investment proposal based only on potential non-financial benefits.	1	2	3	4	5

Q12 Would you consider participating in some interviews after analysing the questionnaire results? 1☐ Yes 2☐ No

Q13 Would you like a copy of the summarised results of the questionnaire?

1☐ Yes 2☐ No

Q14 This survey is anonymous, however information about yourself and your company would be useful when analysing the questionnaire results:

a. Respondent information:

- Your name —————— Telephone ——————
- Your job title ——————
- Number of years you have been
  - in this job ——————
  - in this company ——————
- Your highest degree or qualification ——— Year of award ——
- Your background
  1☐ Accounting and Finance
  2☐ Engineering
  3☐ Operations
  4☐ Mathematics
  5☐ Science
  6☐ Social science/Humanities
  7☐ Other (please specify) ——————

b. Company information:

- Company name ——————
- Approximate annual capital expenditure in your division factory
  £ ——————
- Your company 1☐ a user of AMT 2☐ a vendor of AMT
  3☐ other (please specify) ——————

Thank you for your kind participation in this survey. Please return this questionnaire in the enclosed self-addressed envelope.

Appendix 4 - The Interview Schedule

Investment in AMT

Interviewee:
Job Title:
Company:
Date:

The term AMT (Advanced Manufacturing Technology) refers to any computerised manufacturing process form such as Computer Numerically Controlled machines (CNC), Robotics, Automated Material Handling (AMH), Computer Aided Design/ Manufacturing (CAD/ CAM), Flexible Manufacturing Systems (FMS), and Computer Integrated Manufacturing (CIM).

a. General

- Why has the firm invested in AMT? (Motivations to invest in AMT)
- What are the processes (stages) of evaluating an investment project (for both AMT projects and other investment projects if there is a difference between them)?
- Which departments usually participate in evaluating an investment project (for both AMT projects and other investment projects if there is a difference between them)?

b. Financial Evaluation

- Does the firm use financial evaluation techniques for evaluating capital investments? If so, what are they and why?
- Does the firm use special techniques for evaluating AMT investments? If so, what are they and why?
- Why does the firm use more than one technique?
- Which one is preferred over others? Why this one? Why not the others?
- When PB method is used, how does the firm determine the required PB period? Alternatively, what is the basis for determing the required payback period?

- When NPV or IRR is used, how does the firm determine the discount rate or the minimum required rate?

c. Non-financial criteria

- What sort of non-financial criteria your firm usually use? Why?
- Are non-financial criteria documented in the report of AMT investments evaluation (is there a formal 'non-financial evaluation'?)? If so, How?
- When the firm uses financial and non-financial criteria to evaluate AMT investments, does the firm use scoring method to arrive at a final decision (or recommendation) especially if the results differ between the two categories?

d. Risk

- What sort of risk did the firm expect from AMT investment project? Alternatively, what factors affect the riskiness of an AMT investment project?
- How does the firm consider risk factors? Alternatively, which techniques does the firm use to consider the risk of an AMT investment project? Why?

Index